LONE STAR WILDFLOWERS

LASHARA J. NIELAND   WILLA F. FINLEY

LONE STAR
Wildflowers

A GUIDE TO TEXAS FLOWERING PLANTS

TEXAS TECH UNIVERSITY PRESS

This book is typeset in Stempel Schneidler Std. The paper used in this book meets
the minimum requirements of ANSI/NISO Z39.48-1992 (R1997). ∞

Library of Congress Cataloging-in-Publication Data
Nieland, LaShara J.
Lone star wildflowers : a guide to Texas flowering plants / LaShara J. Nieland,
Willa F. Finley.
p. cm.
Summary: "In photographs and text, describes hundreds of Texas wildflowers.
The 400 photographs are arranged by color to aid identification. The book
describes past and present uses of the plants, the stories behind their scientific
and common names, their medicinal and toxic properties, Native American lore,
and other interesting facts and stories"—Provided by publisher.
Includes bibliographical references and index.
ISBN 978-0-89672-644-4 (paperbound : alk. paper)
1. Wild flowers—Texas—Identification. I. Finley, Willa F. II. Title.
QK188.N54 2009
582.1309764—dc22   2008037857

Printed in Korea

10 11 12 13 14 15 16 17 18 / 10 9 8 7 6 5 4 3 2

Texas Tech University Press | Box 4103 | Lubbock, Texas 79409-1037 USA
800.832.4042 | ttup@ttu.edu | www.ttup.ttu.edu

To our parents,
Myrtle and Henry Finley, and
Charlie and LaRue Shanks.
To Andy Nieland,
chief chauffeur of wildflower photographers—
he brakes for wildflowers!

ACKNOWLEDGMENTS

Thanks to Noel Parsons, director of Texas Tech University Press, for believing in the concept of our book and for his gracious and kind guidance during the project development.

We are grateful to Barbara Werden, design and production manager, for her help and advice in our photography work, encouraging us to strive for increasing excellence. Her experience in book design and her patience in experimenting with the layout were crucial to the development of a high-quality, user-friendly product.

Thanks to the Texas Tech University Press professionals for their cheerful conviviality as they worked around us during our days of editing photos in their office space. In particular, we appreciate Judith Keeling, Karen Medlin, and Lindsay Starr. Thanks to Cynthia Lindlof for her meticulous and sensitive copyediting of the manuscript.

Willa would like to particularly express her gratitude to Dr. Arthur Elliot, who encouraged her to persevere during her graduate studies in botany at Texas Tech and who has always been a loyal and supportive mentor and friend.

Both of us would like to say a special thank-you to Willa's mother, Myrtle Finley, master chef of chefs, who always plied us with exquisite culinary offerings during our intense marathon plant-keying and writing sessions. We did not waste away!

During the lengthy gestational phase of this book, we were continually encouraged by many friends, colleagues, family members, and hundreds of students.

CONTENTS

The seeds for this work were planted when the authors entered Abilene Christian University in September 1970. They have been colleagues in botanical matters since their first day and first class together, with a common interest in biology and plant science and in instructing botany labs. Willa pursued a career in agriculture with a Ph.D. from the University of Nebraska. LaShara earned an M.S. in biology education from Abilene Christian University. Befittingly for botanists, the course of their lives' evolution converged again almost thirty years later in 1997 in the Odessa, Texas, public school system, where LaShara had taught for a number of years and where Willa interrupted her agricultural career to teach science at the secondary level for four years.

For her biology courses, LaShara designed a wildflower project to help her students appreciate plant ecology and its contribution to the rich history and development of the Southwest. She guided her students in making a collection of local wildflowers, identifying them, and learning some of their historical and current practical uses. Subsequently, LaShara and Willa prepared a number of PowerPoint presentations for classroom instruction, using photographs that they had taken of Texas wildflowers over a number of years. They drew on this body of data and photos as a starting point in preparing this field guide and have augmented it with personal experience, interviews with people who use native plants for food and medicine, and primary and secondary literature research of published works relating to the utilization and identification of wildflowers. The authors took all photographs in the book.

This field guide allows the authors to share their love of native plants with the wider audience of Texas flower enthusiasts, who will gain a deeper understanding and appreciation of the ecological and economic value of Texas wildflowers and will expand their aesthetic enjoyment of the colorful native floral displays both in the wild and in xeriscapes.

Many excellent field guides to Texas flowers focus primarily on plant identification and go into considerable detail about plant structures. This book is not intended to supplant those guides but to augment their contributions by providing a broader range of information that reveals the multifaceted character of Texas plants. Included in this book is the following information:

- *Traditional medicinal uses* by Native Americans and early European settlers
- *Food preparations* that are part of the cultures of the Southwest
- *Toxicities* of plants to humans and grazing livestock, noting some of the poisonous chemicals that are responsible for the danger

- *Forage value* for wildlife and livestock
- *Competitive behavior* of plants that have implications for rangeland management or use in landscaping
- *Landscaping suggestions* for using native plants for water conservation, butterfly and bird attraction, and deer resistance
- *Legends and myths* that are based on plant characteristics
- *Origins* of many scientific and common names
- *Growth stages* depicting young plants, buds, mature seed heads, and fruits, in addition to the flowers
- *Seasonal coverage* including spring-, summer-, and fall-blooming species
- *Family descriptions* with typical characteristics, accompanied by representative photos
- *Color coding* to allow more rapid and accurate identification

The book offers additional unique features that set it apart from other wildflower field guides on the market:

- *Accessible format.* The flowers are grouped approximately according to the color spectrum—Red, Orange, Yellow, Green, Blue, and Purple, as well as Pink and White—and within each color category they are further grouped by family and then by genus. The gradation between purple and pink is sometimes difficult to distinguish in the field, and we have therefore positioned the pink flowers after the purple.
- *Reproductive stages.* Photos and text describe the buds and fruits in addition to the flowers.
- *"Exploring Further" pages.* Supplemental photos and text accompany each color category. These provide further details of leaf shapes and arrangements, seedpods and fruits, and both field appearance of plants and close-ups of flowers.
- *Embedded glossary.* Technical terms are explained in context so that the terminology does not hinder understanding and enjoyment.
- *Field-tested material.* Developmental material for this book has been used by students and their parents for over fifteen years. Their suggestions have been helpful in developing this book's user-friendly design.

Many of the folk uses of the plants as medicines, remedies, or treatments for illnesses or medical conditions are anecdotal and unproven by modern medical testing. Because some of these remedies might have adverse effects for some people, they should be used with caution and under the eye of a physician. Furthermore, many plants are poisonous, and mistaking one species for another could have harmful or lethal consequences. The information presented in this book is not meant to encourage experimentation with plants for the purpose of food, medicine, or recreational drugs.

Whether you are a novice to the world of wildflowers or a lifelong botanical enthusiast, this book will enhance your personal appreciation of the beauty and bounty of Texas plants.

LONE STAR WILDFLOWERS

Agave family, Torrey yucca

Pigweed family, pigweed

Sumac family, littleleaf sumac

Milkweed family, antelope horns

Sunflower family, brown-eyed susan

Barberry family, agarita

Agavaceae (Agave Family)

Family traits include perennial growth habit; xerophytic habitat; lengthy flowering stalks; and long, thick, linear leaves, often bearing spines. The flower and fruit structures are similar to those of the lily family and the amaryllis family. Several genera, including *Agave*, *Dasylirion*, *Yucca*, and *Hesperaloe*, that were once classified in the latter two families have now been included in the agave family.

Some species are economically valuable, used to make beverages such as pulque and tequila, to make rope and sacking from sisal, and to produce fructose sugar.

Amaranthaceae (Pigweed Family)
Pigweeds are devoid of any hint of beauty, but they do have their own sort of charm. Pigweed flowers are tiny and without petals. The long, narrow flower head is replete with prickly bracts. Many members of this family have red stems and leaf veins. Some species have been domesticated and are valued for the high protein content of their seeds and leaves.

Anacardiaceae (Sumac Family)
Plants typically bear clusters of small, cream-colored flowers that mature into reddish fruit. The leaves are also colorful, usually reddish or bright yellow when senescing in the fall (*Rhus*). Some species have compound leaves, while others have large, leathery, ovate leaves. The stems contain pungent resin. Economically important members include the gloriously delicious mango, the tasty pistachio, and the crunchy cashew, as well as handsome landscaping plants, such as the smoke tree (*Cotinus*). The family also has a few black sheep, including poison oak and poison ivy.

Asclepiadaceae (Milkweed Family)
The defining characteristic of the milkweed family is the milky sap that comes out of the plant's broken stems or leaves. This milky latex contains physiologically active compounds, particularly cardiac glycosides, which are responsible for milkweed's many medical virtues as well as its toxic properties (it's all in the dosage). The flower structure of milkweeds is so extraordinary and different from the norm that a distinct vocabulary has been devised to describe it.

Asteraceae (Sunflower Family)
The sunflower family is one of the two largest and most diverse plant families in the world; estimates of the number of species range from thirty thousand to sixty-five thousand. For convenience, botanists have divided this huge family into sixteen to twenty tribes based on characteristics of their various flower parts.
The "flower" of a sunflower is actually an extraordinary spiral arrangement of numerous small flowers positioned on a surface appropriately called a receptacle. The flowers are arranged in a Fibonacci series, a sequence of mathematical expressions describing growth patterns within such botanically diverse structures as flowers, pine cones, and cactus spines. In a sunflower, this intricate design of overlapping spirals allows an optimum number of flowers (and then seeds, of course) to be positioned on the receptacle without any of them becoming crowded or stunted by the growth of a neighboring flower.

In a typical flower head, the flowers of the outer ring form a fringe around the edge; each of these ray flowers has its petals conjoined to form what appears to be a single petal (a "ligule"), which is thrust to one side. The flowers in the center, known as disk flowers, have tiny tubular corollas with five tiny "lobes." In keeping with the variety of life, some tribes have only disk flowers, while others have all ray flowers. The stigma typically is bifurcated (two pronged) and is visible above the petals.

Berberidaceae (Barberry Family)

The barberry family is a small one, with only nine genera worldwide and just two in Texas—the genus *Berberis*, or barberry, and the genus *Podophyllum*, known as mayapple or mandrake. *Berberis* is a shrub with stiff, prickly leaves and yellow flowers. The *Podophyllum* genus is friendlier in appearance and is characterized by a herbaceous growth habit and white- to rose-colored blooms. Both genera produce edible fruits.

Bignoniaceae (Catalpa Family)

The catalpa family's shrubs and woody vines offer spectacular floral displays—from the brilliant orange-red of the trumpet vine to the warm pink of the desert willow in temperate regions. Other famous and beautiful members of this family include the purple jacaranda trees in the tropics, the cape honeysuckle and the sausage tree from southern Africa, and the calabash tree from the American tropics. The large, bilaterally symmetrical flowers have a lengthy tube, and the seedpod that follows is long and beanlike. The leaves are usually pinnately compound and opposite each other on the stem.

Boraginaceae (Forget-me-not Family)

Borages have clusters of small yellow, white, or blue flowers, often growing in a bristly scorpioid, or coiled, spike that resembles the unfurling frond of a fern. The seeds are often small and prickly and cling to fabric, especially socks, shoelaces, and trousers.

Several species are cultivated as ornamentals, such as the blue forget-me-not, or as herbs, such as comfrey. Other species are grown commercially for the nutraceutical oil in their seed that contains gamma-linolenic acid, used to treat diabetes, heart disease, cancer, and premenstrual syndrome.

Cactaceae (Cactus Family)

Cacti are well suited for "the land of little rain," with thick, fleshy stems covered by a waxy coating; leaves vestigial or diminished to spines to reduce water loss; and a broad, shallow root system for rapidly taking

Catalpa family, desert willow

Forget-me-not family, puccoon

Cactus family, prickly pear

Caper family, clammyweed

in water from rare rain showers. The beauties among the beasts are the flowers, which are usually large, brightly colored, and supplied with an abundance of stamens.

Capparidaceae (Caper Family)

The caper family is closely related to the mustard family and shares some of its characteristics: the flower parts are in multiples of two, and the leaves emit a sulfurlike smell. The leaves are often compound and are sticky to the touch. The pistils of the caper family are so elongated that they give the plant a spidery appearance, especially when the seedpods are waving about on the end of the long "stalk."

Chenopodiaceae (Goosefoot Family)

Many goosefoot family members have leaves that resemble the webbed feet of geese, hence the name *chenopod*, from the Greek root words for "goose" and "foot." Many species grow successfully in alkaline soil, coping with the osmotic problems of saline conditions by excreting the salt onto the surface of their leaves or stems, giving them a flaky, whitish appearance. Goosefoot flowers are small and inconsequential. In Texas we associate chenopods with weedy species, such as lambsquarters and Russian thistle (tumbleweed). However, the economically important sugar beet and spinach are also in this family.

Goosefoot family, lambsquarters

Dayflower or spiderwort family, dayflower

Morning-glory family, heavenly blue morning-glory

Mustard family, bladderpod

Gourd family, buffalo gourd

Ephedra family, mormon tea

Commelinaceae (Dayflower or Spiderwort Family)

The flowers are in clusters and emerge from a keel-like spathe, a clasping modified leaf. Each flower has three petals: in some species they are equal in size, and in others one of the petals is small and scarcely noticeable. Delicate downlike hairs often cover the stamens' filaments. The delicate blue or purple flowers wilt quickly in the heat of the midmorning sun and deliquesce into an unattractive mushy slime. Leaves exhibit parallel venation typical of monocots and are V shaped in cross section, with their bases sheathing the stem. These plants are copiously mucilaginous, and when the leaves are broken, the mucilage strings out like a spider web.

Convolvulaceae (Morning-Glory Family)
One must get up early to enjoy the full glory of these flowers that greet the dawn. All the colors of the Texas flag—red, white, and blue—are represented in the various species, and some show off with a sporty pink. The family name derives from the Latin *convolvere*, meaning to "twine around," and refers not only to the growth habit but also to the flower buds. The stems are famous for wrapping themselves around any available support, such as fences and other plants. The petals in the buds also twist around themselves like a swirled ice-cream cone before they open up to display their funnel-shaped corolla.

Cruciferae (Mustard Family)
Mustards have four petals in the shape of a cross, or crucifix, hence the family name. The flowers are usually yellow but may also be white or reddish. The two-chambered seedpods have a variety of shapes, ranging from long and thin in the *Sisymbrium* genus to the eyeglass-shaped *Dimorphocarpa* and the round *Lesquerella* pods. The mustard family is economically very important and includes the world's third-most important edible oil crop, canola, as well as cabbage, broccoli, cauliflower, brussels sprouts, and mustard greens.

Cucurbitaceae (Gourd Family)
Members of the gourd family are vines that are usually earthbound unless they grow near a fence or tree that gives them opportunity to climb. Their yellow or white flowers are large and showy and may be monoecious (male and female flowers separate but on the same plant) or dioecious (male and female flowers on different plants). All species, whether domesticated or wild, are well adapted to arid climates. Ornamental gourds and the wild desert plant, buffalo gourd, are bitter tasting and foul smelling, and some parts are toxic if used carelessly. The gourd family is better appreciated for its economically important food crops, which include watermelons, cantaloupes, cucumbers, squashes, and pumpkins.

Ephedraceae (Ephedra Family)
Ephedra is a gymnosperm rather than a true flowering plant, so its reproductive structures are tiny male and female cones, which are found on separate plants. The leaves are small and scalelike and often drop off as the plant matures. The shrublike plant doesn't starve, however, because the stems are green and capable of photosynthesis. *Ephedra* species are found in the deserts of both the Old and New Worlds. An Old

Spurge family, snow-on-the-prairie

Bean or pea family, goldenball leadtree

Ocotillo family, ocotillo

Bleeding heart family, scrambled eggs

Gentian family, mountain pink

Geranium family, stork's bill

World species is used for production of the drug ephedrine, which acts as a decongestant and is used to treat respiratory problems.

Euphorbiaceae (Spurge Family)

Spurges have tiny, atypical flowers, featuring petal-like appendages surrounding a cluster of staminate flowers and a single pendulous ovary, which becomes the fruit. The plants have a wide range of growth forms, from small herbs in North America to trees in the tropics. The family is dominated by the huge genus *Euphorbia*, whose members have a milky sap, or latex, containing a number of biologically active compounds. Most species are poisonous, but some can be treated to render them edible. Cassava, for example, is a very important food crop in tropical climates.

Some species produce the industrial products rubber and castor oil, while others, such as poinsettia, are commercially important as ornamental plants.

Fabaceae or Leguminosae (Bean or Pea Family)

Although most families are identified by their flower structure, legume family members are recognized by their compound leaves and legume fruits, or bean pods. Their pods may be long and thin or short and fat, and most will split lengthwise when they are mature. Legumes have three main patterns of flower shape. The most common type, the Papilionoideae group, with fused petals, looks like an old-fashioned bonnet and is exemplified by the Texas bluebonnet. The flowers of the mimosa group, Mimosoideae, look like a little puffball with a million hairs, which are the anthers; goldenball leadtree is typical of this group. The third type of flower group, Caesalpinioideae, has large showy petals that are not fused; the partridge pea is typical of this group.

Many members of the bean family improve the soil because their roots are home to nitrogen-fixing bacteria that take nitrogen from the atmosphere and convert it into a form usable by plants.

Fouquieriaceae (Ocotillo Family)

This tiny, two-genus family is found only in the southwestern United States and in Mexico. The plants are straight out of *Alice in Wonderland*. Ocotillo has long canes waving as high as twenty feet, while its even stranger relative, the boojum tree, looks like a tall, thin, upside-down carrot, complete with many lateral side branches resembling roots. Both species are well adorned with spines and are often mistaken for cacti. After spring rains, leaves appear all along the stems with red or white flowers on the stem tips.

Fumariaceae (Bleeding Heart Family)

The bleeding heart family has several characteristics in common with the poppy family, particularly the fruiting structures. Like that of the poppy family, the sap contains alkaloids, but in the bleeding hearts it is watery rather than milky. The flower parts of this family are in multiples of two and typically have petals that are joined into cupped spurs. The leaves are usually highly divided.

Gentianaceae (Gentian Family)

This large family does not have many representatives in Texas, but the fame of the Texas bluebell, for which the prestigious Blue Bell ice cream is named, compensates for that lack. The flowers are usually solitary and cuplike, and the petals are often twisted in the bud. The leaves are typically simple and entire, with opposite or whorled arrangement on the stem.

Geraniaceae (Geranium Family)

The Greek word *geranium* means "crane" and refers to the seedpod, which resembles a long-billed bird's head. The geranium family is herbaceous and prefers cool temperatures, often overwintering as a rosette of leaves close to the ground and sending up flowering structures early in the spring. Flower colors usually range from pink to red, and the floral parts are arranged in multiples of five. The leaves are palmately veined or lobed, or they may be highly divided.

Hydrophyllaceae (Waterleaf Family)

Despite the name's implication, only a few members of the waterleaf family prefer aquatic or marshy habitats. Flowers of this herbaceous group are usually blue to purple and emerge from scorpioid, or coiled, spikes. The individual flowers are bell shaped, with the stamens attached to the petals and often extending well beyond them.

Iridaceae (Iris Family)

Plants in this monocot family have long, flat leaves originating from underground rhizomes, corms, or bulbs. Iris family members have flowers that emerge from spathelike bracts, with floral structures in series of three, although the petals and sepals are usually indistinguishable and are called "tepals" as a compromise. The colors of wild family members are often some shade of blue or purple.

Irises are also commercially important as an early spring flower, and a wide variety of colors have been developed, ranging from white to practically black and almost all colors between.

Krameriaceae (Ratany Family)

Krameria plants may be shrubby or trailing on the ground. The bilaterally symmetrical flowers offer a visual surprise for the careful observer: the five rich purplish-red floral parts that the eye first encounters are not the petals but the sepals. The much smaller petals, also five in number, are uniquely modified so that the upper three are fused in a bannerlike structure (somewhat reminiscent of the flared hood of a cobra), while the lower two are sessile and glandlike. The leaves and stems are covered in soft, grayish pubescence. The fruit resembles a mace, the weapon with spikes on a knob.

Waterleaf family, blue curls

Iris family, blue-eyed grass

Ratany family, range ratany

Mint family, henbit

Lily family, crow poison

Flax family, blue flax

Lamiaceae (Mint Family)

Mint family members are easily recognized by their square stems, opposite or whorled leaf arrangement, and bilaterally symmetrical flowers, which are generally two lipped. Plants in this family contain aromatic oils, making them valuable for scents and perfumes, medicines, and culinary spices. Many are of commercial importance, such as lavender, mint, sage, rosemary, and thyme.

Liliaceae (Lily Family)

The flowers in the lily family have three petals and three sepals that are basically indistinguishable from one another, so to eliminate any confusion, they are called "tepals." The leaves are long, narrow, and basal and originate from underground bulbs. Some species, such as onions and garlic, are important food crops, while others are common springtime ornamentals, such as Easter lilies, daffodils, and tulips.

Linaceae (Flax Family)

The flax family is represented in Texas by the genus *Linum*. Plants are usually herbaceous with simple leaves. The flowers are yellow or blue, with five fragile petals.

The economically valuable *L. usitatissimum*, with its delicate blue flowers, is grown for fiber as well as for the oil contained in its seeds. Flax oils are rich in alpha-linolenic acid, an important omega-3 fatty acid that is essential for humans. The seeds also contain mucilaginous sugars that make them effective as a laxative.

Loasaceae (Stickleaf Family)

Touching any part of a stickleaf plant is likely to leave a person with an unwanted souvenir; the barbed hairs that cover all surfaces of the plant have exceptional clinging capabilities. Leaves of some species have another endearing feature: they sting as well as cling. The flowers are white or yellow, with petals in series of five or ten. Some types of stickleaf have only five stamens, whereas others have ten or more. Flowers of all species have radial symmetry.

Malvaceae (Mallow Family)

The mallow family name comes from the Greek word *malakos* for "soft," in reference to the soothing effects that plants in this family have on skin and mucous membranes. Mallows are readily recognized by three typical characteristics: they have many stamens attached along the style (much like a mascara brush), they have five separate petals, and the leaves are usually palmately veined and lobed. This family includes economically important fiber and food crops, including cotton and okra, and well-known ornamentals, such as hollyhock and hibiscus.

Nyctaginaceae (Four O'clock Family)

Nyctaginaceae is a family of night bloomers, which is exactly what their Greek-origin name means. They are commonly called four o'clocks because most of them open in the late afternoon, and they will still be open early in the morning along with the morning-glories.

Four o'clock family members achieve their attractiveness without any petals; the show is put on by brightly colored sepals. The stamens, which are long and extend beyond the rest of the flower, add their flair, too. The leaves are in pairs, and one of the pair is usually larger than the other.

Stickleaf family, tenpetal blazingstar

Mallow family, caliche globemallow

Four o'clock family, devil's corsage

Evening primrose family, pink showy primrose

Wood-sorrel family, yellow wood-sorrel

Poppy family, rose prickly poppy

Onagraceae (Evening Primrose Family)

Members of the evening primrose family have four very delicate petals. The stigma of most species is **X** shaped, while in others it is rounded. Flower colors range from bright white to pastel pink or lemon-yellow. Most bloom either in the cool of the morning or the evening. This family is a source of gamma-linolenic acid, a nutritionally important oil used to treat health disorders ranging from diabetes to premenstrual syndrome.

Oxalidaceae (Wood-Sorrel Family)

Wood-sorrel family members have palmately compound leaves, usually with three leaflets. The flowers are yellow or lavender. Most are small herbaceous plants, but carambola, or star fruit, is a tropical tree that produces tart, tasty fruits. Plants in this family have a pleasant sweet-sour taste because of the presence of oxalic acid, which can be toxic or at least cause the tongue to go numb if consumed in large quantities.

Papaveraceae (Poppy Family)
The buds of poppies are usually nodding, and the flowers have large, delicate petals surrounding the numerous stamens and single pistil. The seedpods have a typical "hat" or "lid" and contain many small, edible seeds that yield an oil used in soaps and paints. Members of the poppy family have bitter-tasting, colorful sap, which may be yellow, orange, red, or a milky white. The sap of the opium poppy (*Papaver somniferum*) is the source of opium, from which heroin, morphine, and codeine are derived.

Passifloraceae (Passionflower Family)
The blooms of passionflowers, varying in color from greenish-white to rich purple, have an exotic-looking fringed corona inside the ring of five petals and five sepals. Strongly curling tendrils enable this vine to cling to trees, fences, or other objects for support.

Pedaliaceae (Sesame Family)
Flowers of the sesame family closely resemble those of the trumpet vine and the snapdragon families: they are long, tubular, bilaterally symmetrical, and brightly colorful. Plants are usually herbaceous, with simple and entire leaves. The family includes the valuable sesame plant that is grown mainly in Southeast Asia for its seed, which is used as a food ingredient and from which oil is extracted.

Plantaginaceae (Plantain Family)
The small, translucent plantain flowers grow in clusters along the flowering stalk. The plants have little or no stem, so the soft, fuzzy leaves are basal, emerging from ground level. Although the members of this family have no beauty to recommend them, they are nonetheless valuable, and some species are grown commercially for their seeds, which are used in laxative preparations. Don't confuse this plantain with the one that is in the banana family.

Polemoniaceae (Phlox Family)
Phlox flower parts are in multiples of five. The five petals are joined at the base to form a tube. In the center of the flowers there is often an "eye" of a different color. The petals in the bud are twisted into a spiral form. Several phlox species have become popular ornamentals for home gardens.

Passionflower family, passionflower

Sesame family, devil's claw

Plantain family, Heller's plantain

Phlox family, blue gilia

Milkwort family, white milkwort

Buckwheat family, smartweed

Polygalaceae (Milkwort Family)

Although this family is widely distributed throughout the tropics and temperate regions, it is represented in Texas only by the genus *Polygala*. Leaves are usually opposite each other on the stem, and flowers are arranged in spikes or panicles. Contrary to the impression given by its name, milkwort does not have milky sap like that of the milkweed family.

Polygonaceae (Buckwheat Family)

The Polygonaceae family is named for its swollen nodes that resemble knobby arthritic knees. Its name is taken from the Greek and means "many knees" or "many joints." These nodes, or places where leaves attach, are usually sheathed by thin membranes called ocreae. The flowers are arranged in nodding spikes, and the fruit is an achene—a miniature version of a sunflower seed.

Members of this family, which includes the cultivated rhubarb species, have an acidic, sour-tasting juice. Their favorite habitat is low-lying, moist areas such as ditches and playa lakes (shallow ephemeral lakes in West Texas and the Panhandle).

Portulacaceae (Purslane Family)

Succulent, fleshy leaves are characteristic of the purslane family. The flowers are solitary, often yellow, pink, red, or white, although what appear to be petals are actually sepals. The fertilized flowers form capsules that split and spill out dozens of tiny silvery-black seeds.

Ranunculaceae (Buttercup Family)

The buttercup family characteristically has leaves that are usually highly divided and plant growth that is variable, from perennial vines sprawling along a fence to tall, robust annuals. The flowers have a profusion of stamens and usually a large number of pistils. Flowers with spurs are common, such as in larkspur and columbine.

Rhamnaceae (Buckthorn Family)

The buckthorn family has members throughout the warmer parts of the world, many of which are armed with large thorns that give the family its common name. The flowers are generally small and unremarkable, with greenish petals. The fruit is a one- to three-seeded drupe—a fleshy structure with a hard inner layer enclosing the seed.

Rosaceae (Rose Family)

Rose family members have radially symmetrical flowers, usually with five petals and the same number of sepals. The numerous stamens yield pollen that is much sought after by insects. Large, leaflike stipules are present at the node where the leaf attaches to the stem. Commercially important members of this family include roses, apples, peaches, strawberries, almonds, and cherries.

Sapindaceae (Soapberry Family)

Plants in the soapberry family are trees, shrubs, or vines that prefer tropical or warm temperate conditions. Flowers have four or five sepals and petals, with the petals often containing nectaries, and twice as many stamens as petals. The flowers are usually dioecious, with male and female flowers found on separate plants.

Purslane family, shaggy portulaca

Buttercup family, columbine

Buckthorn family, lotebush

Rose family, apache plume

Soapberry family, soapberry

Snapdragon family, beardtongue

The genus *Sapindus*, from which this family takes its common name, "soapberry," is valued as a shade tree in gardens and parks. Its fruits are used to make a soaplike substance in tropical countries, but the seed is poisonous. The delicious Asian fruit, lychee, is in this family.

Scrophulariaceae (Snapdragon Family)

Snapdragons have bilaterally symmetrical flowers clustered along a flowering stalk; the genus *Verbascum* is an exception, with radially symmetrical flowers. The tubular flowers have an upper and a lower lip, and the stamens and stigma arrangement varies among the species. Snapdragon flowers are conspicuously colorful and are attractive not only to humans but also to insect pollinators, for whom they have several clever adaptations to facilitate pollen transfer.

Solanaceae (Nightshade Family)

The nightshade family is easy to recognize by its flower and fruit characteristics. In a number of species, the five petals are joined to make a star, and the stamens are shamelessly huge. In others, such as tobacco, the corollas are tubular. The fruit may be a two-chambered tomato-like berry, or a capsule, such as that found in petunia.

The nightshade family is a study in contrasts. On the one hand, its members include very important foods such as tomatoes, potatoes, and peppers. On the other hand, several species, such as tobacco and silverleaf nightshade, are potently poisonous. Then there are others that are simply lovely, such as the commercially important ornamental petunia.

Tamaricaceae (Tamarisk Family)

Tamarisks are shrubs or small trees that grow in deserts and salty areas. They are aggressive invaders along salt pans and streams throughout western Texas and along the Pecos River and the Rio Grande. The leaves are small and scalelike. The tiny pink flowers grow in clusters of dense spikes.

Typhaceae (Cattail Family)

Cattail flowers are unisexual and are borne in dense spikes, the male cluster being directly above the female cluster. The erect leaves often reach ten feet in height, and like all monocots, have parallel venation. Plants have thick rhizomes beneath the mud. The cattail family's preferred habitat is standing water, and plants may become a nuisance by blocking waterways and reducing the holding capacity of bodies of water.

Umbelliferae (Parsley Family)

The flowers in the parsley family look like clusters of white or yellow snowflakes; they are small and arranged in simple or complex umbels, with the flowers' pedicels (short stems) all originating from the same point (like an umbrella). The stems are usually hollow, and the leaves are highly divided pinnate or palmate in arrangement. The family is economically important, with a number of species contributing flavorings, vegetables, and medicines. Many culinary herbs are in the parsley family, including anise, caraway, coriander, dill, and fennel: the seeds are usually ribbed and have oil glands that deliver the characteristic flavor of each of

Nightshade family, silverleaf nightshade

Tamarisk family, salt cedar

Cattail family, cattail

Parsley family, Queen Anne's lace

Verbena family, prairie verbena

Violet family, violet

these species. Other important members of this family include carrots and parsnips.

Verbenaceae (Verbena Family)

In contrast to plant families that are a veritable repository of poisons, verbenas have a user-friendly medicine chest and are pretty besides. Verbena flowers have five two-lobed petals, with one petal slightly larger than the others, giving each individual flower the look of a gingerbread man. This characteristic will help distinguish verbena family members from those in the mint family when one begins to despair about the similarity between their opposite leaves and square stems. They have a long history of use as treatments for an innumerable assortment of ailments,

Mistletoe family, mistletoe

Caltrop family, goathead

and since ancient times they have been associated with magic in religious ceremonies.

Violaceae (Violet Family)

Violet flowers are usually a shade of blue or purple and are bilaterally symmetrical, with flower parts in multiples of five. One of the lower petals is often spurred. Plants reproduce by seeds as well as vegetatively by rhizomes and rootstock division. The pansy, a commercially popular bedding plant, is an important member of the violet family.

Viscaceae (Mistletoe Family)

Mistletoe is a semiparasitic plant that inhabits the upper branches of several species of host trees, including oaks, mesquites, and other deciduous trees, but it remains green year-round after the host tree has shed its leaves. The flowers are tiny, and male and female flowers are separate, some on different plants (dioecious) and some living apart on the same plant (monoecious). The history of mistletoe is rich with magical myths and beliefs, such as the belief that it brings good luck to kiss your sweetheart under a sprig at Christmastime.

Zygophyllaceae (Caltrop Family)

Caltrops are herbaceous plants or shrubs with pinnate leaves that are opposite each other on the stem. Flowers are radially symmetrical and usually have five separate petals. Goathead, or puncturevine, with heavy spines on its seeds, is an infamous member of the family. The well-known wood preservative creosote is extracted from creosote bush, a small, pungent shrub that grows throughout the Chihuahuan Desert.

Red Flowers

ASTERACEAE (Sunflower Family)

Gaillardia pulchella

INDIAN BLANKET, FIREWHEEL

The genus *Gaillardia* was named for the well-to-do French lawmaker of the 1700s Gaillard de Charentonneu, who funded botanical research and exploration. The name "Indian blanket" comes from an old legend about a blanket weaver who wished to thank the Great Spirit for his many gifts. The old man used the colors of the sunset in weaving the most exquisite blanket he had ever made. At his request, he was buried in this yellow, red, and brown blanket so that he could present it as a gift to his Maker. In return, the Great Spirit covered the land with this flower, which displays all the colors of the sunset, just like the blanket.

An ancient Aztec legend tells a different tale of how this flower came to be. According to this story, the Indian blanket was once pure yellow before the coming of Hernán Cortés, but the center became stained red with the spilled blood of the Native Americans he conquered.

This famous drought-tolerant annual is now a well-known ornamental in gardens throughout the United States, although the larger and more robust *G. grandiflora* is showier and longer lasting.

Kiowas thought this flower would bring them good luck and kept it in their homes. Many tribes used a tea from this plant as a diuretic to relieve water retention and to treat urinary problems. Pale green and bright yellow dyes can be extracted from the flower heads, depending on the compounds used to fix the colors.

Gaillardia suavis

FRAGRANT GAILLARDIA

Fragrant gaillardia may be distinguished from its cousins by its pleasant aroma, which is given recognition through its name, *suavis*, which means "sweet." Like all members of this genus, fragrant gaillardia has a brownish-red sphere of disk flowers, but this species usually has no ray flowers, or if they are present, they are short and stubby.

The Blackfeet made a tea from the roots of various *Gaillardia* species for gastroenteritis, and a powdered form of the root was used to treat skin diseases. From the entire plant, the Blackfeet also made a wash to soothe irritated nipples of nursing mothers as well as sore eyes and noses.

BIGNONIACEAE (Catalpa Family)

Campsis radicans

TRUMPET VINE, TRUMPET CREEPER

The woody vine *C. radicans* has large orange-red blossoms that are tempting for hummingbirds and bees. This native of East Texas is used throughout the state as an ornamental. It climbs up any available supporting structure and has large, showy blooms throughout the summer. It is easily propagated by seed or cuttings.

Trumpet vine has antifungal properties and has been used as a soaking agent to heal athlete's foot and toenail fungus. Yeast infections have been treated with a wash made from the leaves and the flower. People

Trumpet vine

with sensitive skin should exercise care when handling trumpet vine since it may cause a rash.

CACTACEAE (Cactus Family)

Coryphantha macromeris var. *runyonii*
(*C. runyonii* or *Mammillaria runyonii*)

RUNYON'S NIPPLE CACTUS, DUMPLING CACTUS

This cactus is named for the resemblance of the tubercles to nipples of mammary glands. There are about three hundred species of this particular genus, with most being native to Mexico. Runyon's nipple cactus, found in far South Texas, was named in honor of Robert Runyon, a botanist from Brownsville. In addition to serving as a visual feast to all passersby, the pinkish red flowers of this six-inch-tall cactus advertise ample offerings of nectar and pollen for bees and other pollinators.

The Apache, Navajo, Tewa, and Pima peoples consumed its fruits fresh, dried, or cooked. The Tewas burned the spines off the stems and ate the entire plant. The Pimas have used this plant to treat earaches.

Echinocereus coccineus (E. triglochidiatus)

CLARET CUP CACTUS, HEDGEHOG CACTUS

One of the most breathtaking cacti is the claret cup cactus when it is adorned with its large red blooms. The entire plant is pleasing in its shape, but the bright red flowers are what really set it apart. Claret cup cacti form large clumps that can reach up to four feet in diameter. It is an excellent choice for a xeriscape.

Members of the Tarahumara tribe in southern Texas and northern Mexico have used its toxic alkaloids as a "false peyote" in certain hallucination-seeking rituals.

A view of the whole plant is found on p. 34.

Opuntia macrorhiza var. *pottsii*

RED PLAINS PRICKLY PEAR

The red plains prickly pear, a relatively small cactus with very long spines, has a red-orange blossom. This species is found from the Panhandle to South Texas.

Like other types of prickly pears, this species has edible pads and fruits, and the juicy pulp may be used for a hair conditioner. The juice has soothing effects on the urinary tract and on mouth and skin sores.

Prickly pears also have other flower colors. For an additional photo, see p. 81.

FOUQUIERIACEAE (Ocotillo Family)

Fouquieria splendens

OCOTILLO, DEVIL'S WALKING STICK, CANDLEWOOD

Ocotillo was named for the nineteenth-century French botanist Pierre Fouquier. It lives up to its species name, *splendens*, when the tips of its tall stems blossom with splendid inflorescences of flaming red flowers. These waxy, tubular blooms attract a host of pollinators, including humming-

birds, finches, and orioles, as well as bees and other insects. After pollination, the flowers are succeeded by orange berries, which are notable in their own right. Ocotillo has several more-or-less erect woody stems, often exceeding twenty-five feet in height, which add a strong vertical element to a xeriscape.

Some people mistakenly consider the desert-dwelling ocotillo to be a cactus, since its stems are so very spiny, but it is in a different family altogether.

Stems with the flowers attached can be boiled to make a tea for treating sore throats, bladder infections, and prostate problems. Bark tea is used to loosen lung congestion. The crushed flowers and roots are added to a steaming hot bath for a fatigue-relieving soak. Extracts from the flowers and roots are used to arrest bleeding from superficial wounds.

This plant has a novel adaptation to life in an arid land: when there has been no rain, it will remain dormant, even to the point of shedding its leaves. The stem contains a leaf-inhibiting chemical that must be washed away by rain before the foliage will be able to emerge. This strategy assures the plant of effective utilization of scarce moisture.

While the plant is dormant, it may be cut and stuck in the ground, where it will take root and grow. A series of such stems may be used to create "living fences."

Close-ups of the leaves and flower spike are found on p. 34.

MALVACEAE (Mallow Family)

Malvaviscus arboreus var. *drummondii* (*M. drummondii*)

TURK'S CAP, MEXICAN APPLE, MAYAPPLE

One of Texas' best-loved and most-used flowering ornamentals, turk's cap has unusual blooms that feature a well-extended style and stigma surrounded by a twist of red petals. This unique flower is the inspiration for its imaginative common name, while its genus name, *Malvaviscus*, means "sticky mallow."

In its native habitat from the Edwards Plateau to Mexico and Florida, turk's cap is a fairly large and coarse-looking multibranched shrub, sometimes reaching up to nine feet in height, although usually it is much shorter. Although it is drought tolerant as far west as the Midland/Odessa area, turk's cap thrives in more humid regions.

Turk's cap's leaves have been used as a soothing, skin-softening emollient, and in Mexico the flowers are used to make a tea to treat inflammation of the digestive tract, to relieve menstrual discomfort, and to stop diarrhea. A poultice of the leaves and roots may be applied to the chest to ease pulmonary congestion.

The marble-sized red fruit is edible for humans as well as for a variety of birds and other animals. It is said to taste something like a watermelon or apple. It makes a flavorful amber-colored jelly. A tasty tea may be brewed from the dried flowers, and the leaves may be cooked with other greens. The leaves and flowers yield dyes that range in color from tan to peach or mauve.

Ruby-throated hummingbirds and several species of butterflies seek out turk's cap flowers for the nectar. Livestock occasionally browse the leaves.

NYCTAGINACEAE (Four O'clock Family)

Nyctaginia capitata
SCARLET MUSKFLOWER, DEVIL'S CORSAGE, DEVIL'S BOUQUET

Nyctaginia means "night blooming," and the scarlet muskflower's lovely red flower clusters open at night and are still at their brilliant best early in the morning. On rare occasions, individuals will have pale pink blooms. Its home is in the western and southern parts of Texas, where it has strong potential for use in xeriscaping, with its ability to grow with little or no supplemental watering.

The plant is called devil's corsage because the malodorous flower head will stick to a shirt like a corsage. In fact, the entire plant is covered with a sticky secretion that causes dirt to adhere to the surface, giving it an untidy look that is out of character with its ravishing color.

Like all members of the four o'clock family, this flower has no petals.

What you see are actually brightly colored sepals. The long, red stamens extend far beyond the flower, giving the cluster a frilly appearance. A close inspection reveals that each anther has two wonderful hot pink pollen sacs, making the bloom even more visually fascinating. Too bad it doesn't smell as good as it looks!

Views of the rare pink flowers and the fruit are found on p. 34.

PAPAVERACEAE (Poppy Family)

Papaver rhoeas

CORN POPPY, FLANDERS POPPY, RED POPPY

The scientific name *Papaver rhoeas* means simply "poppy red." It is commonly called "corn poppy" because it can become a pesky weed in cultivated fields.

Originating in the Old World, corn poppy has escaped cultivation to become naturalized in Texas, much to the delight of flower enthusiasts. Blooming from about April to June, the large blossoms are most frequently an eye-catching red with a black spot at the base of each petal. Occasionally the flowers are purple or even white. The bristly stem, which contains yellow sap, may be up to three feet tall, and the hairy leaves are deeply incised.

Corn poppies make a delightful and hardy addition to a home garden.

Seeds should be planted in the fall, in well-drained soil that receives full sun. Planting poppies in alkaline soil will enhance the development of the red pigments.

Poppy seeds add a nutty flavor to baked products and to salad dressings. The seeds are also the source of an oil that has qualities similar to those of olive oil. The leaves may be eaten raw or cooked as long as they are gathered prior to the formation of flower buds. After that, they are mildly toxic. The flower petals are used to make syrups and beverages, and the red pigments add color to inks, potpourri, and food products, especially wine.

Because poppies are slightly narcotic, they have traditionally been used to relax an agitated person and to aid sleeping. They have also been utilized to relieve pain, quell coughing, and aid digestion.

A view of the whole plant with buds is found on p. 34.

POLEMONIACEAE (Phlox Family)

Ipomopsis rubra

STANDING CYPRESS, SCARLET GILIA, TEXAS PLUME

There could be no more appropriate name for standing cypress than *Ipomopsis rubra*, which means "striking appearing red one." Growing to heights in excess of six feet, the spires of scarlet flowers brighten Texas roadsides through the summer and fall.

The threadlike leaves superficially resemble those of cypress trees and are the source of its common name. Some Native American tribes called standing cypress "chigger weed" because it grows in areas inhabited by those pesky parasites.

The long-tubed red blooms are irresistible to hummingbirds, which hover about sampling the contents. Hawk moths, with their long mouth-parts, come calling at night to have a few drinks.

Elderly people tell of a time when standing cypress covered the hills of Central Texas, but sadly, the population has been much reduced. However, seeds are commercially available, and they are reasonably easy to grow in a wildflower garden. The seeds do not germinate uniformly, so don't be alarmed if some wait to germinate in the second year.

SCROPHULARIACEAE (Snapdragon Family)

Castilleja indivisa

INDIAN PAINTBRUSH

Paintbrush's wide range of "floral" colors is the basis of a popular tale about their origin. According to the legend, they were originally green leafy plants that were dipped in paints by a Native American artist to make a picture of a sunset on a deerskin, and as they were discarded, the plants took root and grew, their leaves still colored by the paints. In modern times, the Texas Department of Transportation has taken over the role of distributing paintbrushes and has sown seeds along Texas highways since the 1930s.

Paintbrushes are great pretenders. Looking for all the world like large blooms, the "petals" are actually leaves putting on a show for the insects. The real flowers are nothing to brag about; they are small, pale green, tubular affairs that get lost among the colorful bracts, which is the botanical term for the colored leaves. Paintbrush species can be distinguished by the degree to which the bracts are divided.

Native Americans found that they could eat the plants, but not in great quantities, because paintbrushes accumulate toxic metals from the soil. They also learned to make a tea with diuretic qualities effective in treating kidney problems. Paintbrush tea is an old home remedy for venereal disease, probably because it helped soothe the pain in the urinary tract that is often symptomatic of STDs. It also served to alleviate stomach discomfort. The whole plant may be steeped in a tub of hot water to soak away one's aches and pains. Crushed paintbrush was part of a traditional remedy for treating leprosy.

With such beauty, it's little wonder that paintbrushes are in demand as garden flowers. However, one should never contemplate digging up established wild plants, not least because paintbrushes do not transplant well. Therefore, one must begin with the seeds, which are very tiny and difficult to handle. They may disappoint a hopeful gardener by not sprouting the first year. Since paintbrushes are semiparasitic, they also need a suitable host plant for their roots in order to become established, and the lack of such a host may be another reason for their failure to appear when expected.

Early settlers, as well as Native Americans, extracted dyes from paintbrushes in tones of greens, yellows, and rusty orange. Minerals were mixed with the roots to make a black pigment for coloring deer hides.

Field views of other paintbrush species are shown are shown on pp. 34 and 35. Note color variations.

In-depth views of fruits and whole-plant views are provided in this section. Take particular note of the profusely hairy fruiting structure of scarlet muskflower, as well as its rare pink flower form.

Echinocereus coccineus, *claret cup cactus; field appearance, p. 25.*

Fouquieria splendens, *ocotillo; leaves appear from thorn bases following a rain and then fall away, p. 26.*

Fouquieria splendens, *ocotillo; flowers borne only at the end of the long canes, p. 26.*

Nyctaginia capitata, *scarlet muskflower; immature anthocarps, each containing a seed and showing an abundance of gland-tipped hairs, p. 30.*

Nyctaginia capitata, *scarlet muskflower; rare pink form, p. 30.*

Castilleja integra, *Indian paintbrush; field view showing orange phenotype, p. 32.*

Papaver rhoeas, *corn poppy; field view showing nodding buds and dense hairiness, p. 31.*

Castilleja purpurea, *Indian paintbrush; field view showing pink phenotype, p. 32.*

Castilleja purpurea, *Indian paintbrush; yellow phenotype, p. 32.*

Orange Flowers

ASCLEPIADACEAE (Milkweed Family)

Asclepias tuberosa
BUTTERFLY-WEED, ORANGE MILKWEED, INMORTAL

The large, tuberous root of butterfly-weed provides the basis for the name *tuberosa*, and the common names are based upon its attractiveness to butterflies, its color, or its medicinal properties.

Butterfly-weed is atypical among milkweeds in that the sap is not milky, but like its relatives, it contains cardiac glycosides that are toxic because of their action on heart and uterine muscles. Some of these compounds also show antibiotic activity and tumor cell inhibition.

The powdered roots have been used to treat inflammation of the lungs, generically called pleurisy, and to bring relief to tuberculosis patients. The active ingredients cause the patient to cough and clear the breathing passages, but when the medicine is ingested in excess amounts, the same compounds may cause nausea and vomiting. The roots have also been used to treat rheumatism, wounds, gastrointestinal illnesses, urinary disorders, and even smallpox. As a testament to the importance and significance of butterfly-weed among the Omaha people, only a few men were authorized to gather and prepare the roots, following special ceremonial procedures.

American Indians used the strong fibers to make bowstrings, and they extracted a red dye for decorating their baskets.

For additional information on milkweeds, see pp. 150, 186, and 233.

ASTERACEAE (Sunflower Family)

Ratibida columnifera (*R. columnaris*)

MEXICAN HAT, PRAIRIE CONEFLOWER

Mexican hat is one of Texas' most readily recognized wildflowers. The image of a tall-crowned, floppy-brimmed sombrero leaps to mind, or dancing girls with yellow or brown skirts fluttering in the wind. Ray-flower colors vary from bright yellow to brown to reddish-maroon—and some show off all three together.

Mexican hat is very eclectic in its soil preference and grows equally well in native prairie or disturbed roadside soils. It is also highly adaptable to the amount of water available and will grow well in a xeriscape or produce prolifically if it is given generous amounts of water. Be prepared to reduce the huge number of seedlings that spring up in the fall, or the merry sombreros will be taking over the garden vistas.

The Dakota people made tea from the leaves and stems for treating stomachache and tea from only the flowers for headache relief. They also used the flower heads and other parts to treat wounds and to stop chest pains, besides enjoying tea from the leaves just for its beverage qualities. The Cheyennes treated snakebite and poison ivy rash with a strong brew made from the leaves and stems. Yellow and green dyes can be made from this plant.

Other colors of Mexican hat are shown in the Exploring Further section, p. 45.

LEGUMINOSAE or Fabaceae (Bean or Pea Family)

Sesbania macrocarpa (S. exaltata or *S. herbacea)*
TALL INDIGO, RIVER HEMP, PEA TREE

Standing up to twelve feet tall, *Sesbania* has fernlike compound leaves and beautiful small flowers that vary in color from yellow to orange. It prefers the moist soil of lakeshores, river sandbars, and marshes. It can become a weed in cultivated areas such as rice paddies.

Like all legumes, *Sesbania* produces seeds in bean pods. These seeds contain saponins, or bitter glycosides, which are poisonous to some birds, particularly domesticated chickens. Feed that is accidentally contaminated with *Sesbania* seeds will result in lower weight in broilers and reduced egg production. In light of that, it is somewhat surprising that these seeds are eaten with impunity by quail, dove, and turkey and are even marketed as bird feed.

Sesbania has also been suggested for use as an intercrop in pecan and citrus orchards. Not only will it provide natural fertilizer, due to the nitrogen-fixing bacteria it harbors in its root system, but also it has the potential to help control harmful insects by attracting them away from the trees. Other species in the genus *Sesbania* are used for forage for livestock and for many herbal medicines, particularly in Asia and Africa.

The tall stalks of this annual are the source of strong, decay-resistant fibers that were used by the Yumas to make nets and fishing lines. A view of the whole plant is found on p. 45.

MALVACEAE (Mallow Family)

Sphaeralcea angustifolia

COPPER MALLOW, COPPER GLOBEMALLOW

A beginning wildflower enthusiast sometimes has difficulty in distinguishing copper mallow from caliche globemallow. Both have almost identical orange flowers, but the main difference lies in the leaves. Copper mallow has long, narrow leaves that are not divided into lobes like those of caliche globemallow, and it is usually taller.

The ancient people who inhabited the Four Corners area (where Arizona, Utah, New Mexico, and Colorado join), as well as their descendants today, have made use of globemallow for medicine. They treat stomach discomfort, coughs, and infections with this plant. They also make a diuretic tea from mallows to alleviate water retention and to treat urinary tract problems. The boiled roots relieve constipation. The Navajo people consider it to be one of the Life Medicines, indicating its importance in their culture.

During lean times Native Americans ate mallows as food, but this was not generally done except when little else was available. They also extracted a pale orange dye from some species.

Sphaeralcea coccinea

CALICHE GLOBEMALLOW, YERBA DE LA NEGRITA, SORE EYE POPPY

Caliche globemallow offers cheerful color on the caliche outcrops and gravelly areas where it thrives.

Like virtually every other member of the mallow family, caliche globemallow serves as a demulcent and emollient, softening skin and relieving irritation. Native Americans were well aware of the soothing properties of this plant. The Hopis called the caliche globemallow "sore eye poppy" for its value in treating eye irritations, although the fine leaf hairs are themselves said to be very irritating to the eyes, causing red, itching "sore eyes." Babies were given the root to chew to assuage teething pain. The leaves were crushed to obtain the slightly slimy juice used to treat skin irritations. A good trick that hikers should keep in mind is to put leaves in their shoes to appease sore, blistered feet. Not only do the leaves cushion the foot but with each step the soothing sap is released on the irritated skin.

The beneficial effects of mallow juice can also work internally. Leaves and flowers can be chewed or brewed for relief of sore throats or for minor digestive upsets. Native Americans who suffered from the pain of hemorrhoids would roll their own suppositories with mallow leaves and a little tobacco moistened with saliva. Caliche globemallow root was used to relieve the pain of bladder infection. Native American women of the Upper Great Plains made a tea from a mixture of the root of caliche globemallow and gum from chokecherry stems to treat postpartum hemorrhaging.

Native Americans boiled this plant to make a hair gel. If they decided to give their "do" softness and shine, they worked the gel in and then rinsed it out. However, if a stiffening agent was called for, the gel was left in as the hair dried. Clearly, the desire to be fashionable hasn't changed, just the means of achieving "the look."

PAPAVERACEAE (Poppy Family)

Eschscholzia mexicana

MEXICAN GOLD POPPY

Mexican gold poppy is found in the Trans-Pecos region, where it makes a bright orange show from March to May. This many-branched plant usually grows about one foot high and has lacy, blue-green foliage. The numerous seeds ensure that this prized plant will come back year after year. This species, like its cousin California poppy, is a brilliant choice for a wildflower garden.

Like all poppies, this species has edible seeds, but the plant as a whole should be considered inedible because of the toxic alkaloids in its yellow sap. The effect of the alkaloids is much weaker than those of opium poppy, and the plants are useful as pain relievers and sedatives. A tea brewed from the entire plant treats bladder and prostate pain and also serves as an analgesic and antibacterial wash for minor burns and scrapes. The sap has long been touted as an effective wart remover.

Mexican poppy accumulates copper in its plant structures and is used as an indicator plant to identify soils rich in copper.

A view of the whole plant with buds and fruits is found on p. 45.

This section provides photographs of color variations, fruits, and whole-plant views.

Ratibida columnifera, *Mexican hat; color variant, p. 39.*

Ratibida columnifera, *Mexican hat; color variant, p. 39.*

Ratibida columnifera, *Mexican hat; color variant, p. 39.*

Ratibida columnifera, *Mexican hat; color variant, p. 39.*

Eschscholzia mexicana, *Mexican gold poppy; note that seedpods are longer and narrower than buds, p. 43.*

Sesbania macrocarpa, *sesbania; field appearance showing the height (up to twelve feet) and the pinnately compound leaves, p. 40.*

Yellow Flowers

ASTERACEAE (Sunflower Family)

Amblyolepis setigera

HUISACHE DAISY, BUTTERFLY DAISY, HONEY DAISY

Huisache daisy is an early herald of spring and one of the first to bloom. In South Texas it is often found growing under huisache trees so has taken on the common name of its much taller friend. In other areas it grows without the benefit of trees, usually in large tribal units that are arrestingly conspicuous in brilliant yellow.

Huisache daisies have a strong, sweet aroma that comes from the presence of coumarin, a compound that slows blood clotting. Livestock relish huisache daisy, but ranchers should limit access to the plant in order to avoid coumarin poisoning.

The size of the flower varies considerably with the availability of water. In a rainy year, or in one's home garden, the flowers become quite large and especially delicious, both to the human eye and nose, as well as to the taste buds of butterflies.

Berlandiera lyrata
CHOCOLATE DAISY, SOFT GREEN EYES

Unlike other lackadaisical daisies, this one is up early, brightening the morning with cheery yellow petals and delicious chocolate aromas. As the day heats up, chocolate daisies' petals begin to shrivel, and they are practically unnoticeable by day's end. After the flower matures to seed, a velvety green base remains that resembles its other moniker, "green eyes."

With its heady aroma of chocolate, it is only fitting that it was named for the Swiss-French explorer Jean Louis Berlandier, who at the tender age of twenty-one was commissioned as a botanist to accompany an expedition to Mexico, where he settled and became a medical doctor, as botanists were wont to do with their skills in the art of healing with plants and herbs.

Native Americans used chocolate daisies to season food, especially meat dishes.

A view of the mature seed head is found on p. 111.

Coreopsis tinctoria
COREOPSIS

Coreopsis seeds look like small, skinny bugs and earned this plant its genus name: *core* from a Greek root word for "bug"; and *opsis*, which indicates resemblance. There is yet another bug connection: early pioneers discovered that fleas and bedbugs could be repelled by putting dried coreopsis flowers into their mattresses.

The species name *tinctoria* refers to a stain or dye, and the entire coreopsis plant can be used to make lightfast dyes ranging in color from rusty red to yellow or green.

The bright yellow and red flowers have led to this species' domestication for the home garden, and several color variations have been identified. Springtime and early summer find the eastern half of the Texas countryside painted with acres of wild coreopsis blooms. In the more arid lands of West Texas, coreopsis is restricted to areas of moist soil, such as the edges of playa lakes.

Coreopsis

Native Americans used the yellow flowers to make a red-colored beverage tea. The tea served other purposes as well, including relief for congestion due to colds and for internal pains and bleeding.

Dracopis amplexicaulis

CLASPING-LEAF CONEFLOWER

The flowers of clasping-leaf coneflower are deceptively similar to those of Mexican hat, but an examination of the leaves will immediately eliminate any confusion: note how the base of the leaf clasps the stem. This botanical embrace is in fact the basis for its species name: *amplexicaulis* means "clasping the stem."

Clasping-leaf coneflower reaches two or more feet in height, and its bright golden and brown colors fill ditches, well-watered fields, and even a garden if it is (foolishly) invited in. It reseeds itself abundantly, and if it is in a small garden area, one will need to be very proactive in removing the extraneous seedlings. It grows especially well in areas with clay soil or poor drainage. Unfortunately, clasping-leaf coneflower is very suscep-

tible to powdery mildew, and after a few days of cool nights followed by hot sun, it will become infested with this white fungus.

Dyssodia acerosa

DOGWEED, FETID MARIGOLD

The rich color of the dogweed petals belies the unpleasant smell of the crushed leaves. Its genus name is aptly based on the Greek word *dyssodia*, which means "stench." The numerous small oil glands on the leaves are the source of the pungency. Actually, it smells something like marigold, which isn't all that bad to some people. Dogweed can grow to one foot or more in height, with many branches, and can be a perennial in warmer regions.

Dogweed's name refers to its original habitat around the disturbed soil of prairie dog towns. These days, with no shortage of disturbed roadsides and road cuts, dogweed has successfully expanded its territory.

In keeping with the spirit of fighting one olfactory assault with another, the Omaha and Sioux peoples used powdered dogweed as snuff to cause nosebleed and thereby relieve headaches.

A closely related species, *D. pentachaeta* (parralena), is shown on p. 111.

Engelmannia peristenia (E. pinnatifida)

ENGELMANN DAISY, CUT-LEAF DAISY

This genus was named in honor of Dr. George Engelmann, a physician-botanist who collected flowers in Texas during the mid-1800s. Engelmann daisy is chic and charming enough to earn a place in the home garden, and unlike some other wildflower species, it knows how to be neighborly and doesn't crowd out other species.

Livestock appreciate this tasty daisy because of its high protein content, estimated to be almost 30 percent of dry matter. Thus, it is more likely to be found on roadsides or in the home garden than in grazed pastures.

The entire plant, including flowers, can be used to make a yellow dye. A view of the whole plant is found on p. 111.

Gaillardia pinnatifida

YELLOW GAILLARDIA, CORONILLA, YERBA DEL SOL

Like its cousins with the same genus name, yellow gaillardia was named for Gaillard de Charentonneu, the wealthy French lawmaker of the 1700s who funded botanical research.

Native American women believed that their fertility would be improved if they drank a tea made from yellow gaillardia for seven days in a row. The tea was also used as a diuretic to alleviate water retention and to relieve the pain of bladder infections. An old folk remedy for sinus headaches involved making a poultice of the entire plant and applying it

to the forehead. Inhaling the powdered flowers was also used to relieve headaches—after the patient finished sneezing!

Yellow gaillardia contains alkaloids that may cause liver damage if used over an extended period of time.

Grindelia ciliata (*Prionopsis ciliata* or *G. papposa*)
SAWLEAF DAISY, GUMWEED, WAX GOLDENWEED

As its name indicates, sawleaf daisy's leaf edges have protective spines all along the margins, effectively preventing its consumption by livestock. This stiffly erect plant with bright yellow blooms may reach six to seven feet in height and blooms from late summer into fall.

Native Americans collected sap from broken flower stems and used it for chewing gum, just as they did with its cousin, curly-cup gumweed, with which it shares many medicinal properties. (See the next entry.)

Although some botanists have recommended this flower as a good background plant in a xeriscape, practical experience has proven that it quickly becomes a nuisance because it spreads so easily by seeds.

Close-ups of the leaves and of the stem are found on p. 111.

Grindelia squarrosa

CURLY-CUP GUMWEED, ROSINWEED, TARWEED

Curly-cup gumweed's scientific name was bestowed in honor of David Grindel, an Estonian pharmacist and botanist (1776–1836).

Curly-cup gumweed has compactly attractive yellow flowers, with recurved bracts (phyllaries) under the flower head that give it the first part of its common name. The Dakota Indians had the same idea and called it "curly buffalo." These bracts, and indeed the whole flower bud, are covered with a sticky, balsamic-flavored gum that can be chewed like chicle; in fact, the Indian tribes in the northern Great Plains chewed the entire flower head. This is the source of the other part of its name, "gumweed," which the Blackfeet recognized with their name for it: "sticky weed."

Curly-cup gumweed is well adapted to disturbed areas or overgrazed pastures and is also found on the edges of playa lakes of West Texas and the Panhandle.

The flower heads and leaves of curly-cup gumweed are the primary medicinal agents, with roots used only rarely. Its most important use, for

which it achieved recognition in the *U.S. Pharmacopoeia* in the late 1800s and early 1900s and in the *National Formulary* between 1926 and 1960, was as a treatment for poison ivy rash. A number of North American Indians, as well as early American settlers, made a bitter aromatic tea from the young leaves and flower heads to treat bronchitis, pneumonia, and asthma or to calm the spasms of coughing. The Sioux also used it to treat urinary infections, and the Gros Ventre, Shoshone, and Cree peoples used it for combating venereal diseases.

Yellow to gold dye can be extracted from the flowers, stems, and leaves. The plant's most mundane use is as a broom, a small "by-product credit" after all the more valuable medicinal extracts have been removed.

Gutierrezia spp.
BROOMWEED, SNAKEWEED, MATCHWEED

The broomweed or snakeweed group contains both annual and perennial species that share a number of physical and chemical characteristics as well as common names. The plants have a hemispherical profile, the outer boundary of which is composed of a multitude of small yellow flower clusters. Early European settlers used this plant to make brooms, hence the common name.

Pastures that are severely overgrazed can become covered with broomweed because livestock avoid eating it, allowing it to proliferate. These plants thus serve as an indicator of poor grazing practices.

The resin found in broomweed gives it a pleasant pinelike smell and makes it useful as kindling for fires. The same smell is indicative of toxic saponins that can cause abortions or death in cattle and sheep. Perennial species of broomweed accumulate selenium when growing in soils high in that element.

Like their ancestors, the Pueblo people of the U.S. Southwest still use broomweed for a number of medicinal applications, including treatment for such assorted ailments as stomach disorders, colds, eye problems, snakebite, and insect bites. For traditional arthritis treatment, the tops of the plant are boiled in water, and the resulting tea is divided, some for drinking and the rest for adding to a soaking bath. Broomweed's effectiveness is said to rival that of aspirin for arthritis relief. Kiowa Apaches drank a tea of broomweed for pulmonary problems; the Blackfeet boiled the root and inhaled the steam for respiratory disorders. Pioneers made a syrup from the dried flowers to clear congested lungs. Today, Navajos make a colorfast yellow, gold, or green dye from the plant.

A close-up of the flowers is found on p. 111.

Helenium amarum
BITTERWEED, BITTER SNEEZEWEED

Helenium amarum is an annual species with two varieties distinguished by the color of the disk flowers. In the eastern part of Texas the dominant disk-flower color is yellow, while in the western regions of the state it is brown. The ray flowers in both varieties are a rich lemon-yellow.

Livestock usually avoid eating these weeds because the smell is not inviting and the taste is bitter, with the unpleasant flavor increasing at flowering time. However, if pastures are overstocked with animals so that grazing pressure is too high, the lack of competing plants will allow bitterweed to increase in number and degrade the value of the pasture.

Sheep or cattle that eat bitterweed will produce unpalatable milk or meat and, in extreme cases, experience weakness and digestive problems that may lead to death. Even honey made from its nectar will be bitter.

The sneeze-inducing properties of bitterweed gave it several interesting medicinal uses among the Great Plains and Southwestern Indians. To clear the sinuses and other congestion from colds or flu, people inhaled the powdered flowers, and the resulting sneezes did the rest of the work. Violent sneezing also helped women to clear the afterbirth following the delivery of a baby.

Bitterweed causes allergic reactions in many people, and one should take care when collecting this specimen. One of the allergens, helenalin, shows some antitumor activity.

Another bitterweed, *H. flexuosum*, is a perennial found in the eastern half of the United States, including Texas. Close-ups of the flower head and the winged stems may be seen on p. 112.

Helianthus annuus

COMMON SUNFLOWER

The sunflower has been one of the most useful plants to Native Americans, serving as a source of food and medicine as well as of dyes and fibers.

The seeds provide food for both animals and people. Native Americans ground sunflower seeds into flour and pounded or boiled them to extract the oil. Early pioneers made a coffee substitute from the roasted shells. Some believe the young buds taste like artichokes when boiled and served with butter, but others perceive the flavor to be more akin to paint thinner!

Tea from boiled leaves was used to reduce fever and to heal skin damage from screwworms. Rheumatism and snakebite were treated with a tea made from the roots.

The flowering heads yield dyes in shades of yellow, orange, and green. Hopis are said to be able to extract a shade of blue from the seeds, and they use these dyes to paint their bodies for certain ceremonies.

Many Native American tribes used sunflower oil both for cooking and for a hair dressing. The Pimas and Maricopas made chewing gum, as well as candles, from the pith of the stem. Fibers from the stem can be used to make paper.

Helianthus ciliaris

BLUEWEED

Were blueweed's character judged on its flower alone, it would have a more favorable reputation. Instead, it is known as a noxious perennial weed that aggressively spreads by underground rhizomes, forming huge colonies that are difficult to eradicate. Persistent spraying with appropriate herbicides can effect some control, and long-term grazing pressure by sheep has been found to eliminate blueweed infestations—without apparent harm to the sheep. Furthermore, blueweed is believed to contribute to hayfever woes when it is in bloom.

Blueweed is aptly named for its blue-green foliage and stems. Its leaves are tough and wiry and exude a strongly aromatic smell when crushed.

A view of the whole plant and of the mature seed head are found on p. 112.

Helianthus maximiliani

MAXIMILIAN SUNFLOWER

In the early 1800s a German prince, Maximilian Alexander Philip von Wied-Neuwied, undertook an extensive botanical expedition through the western United States and collected this stately plant. It is well named in his honor.

Maximilian sunflower is the much taller cousin of common sunflower. Both are skilled in the art of self-propagation, although by different means; common sunflower is an annual and reproduces by means of seeds, while Maximilian sunflower not only produces quantities of seeds but also spreads by an extensive underground rhizome system.

Maximilian sunflower grows robustly as long as there is some soil; it doesn't seem to care what kind or where and is able to tolerate a range of moisture availability. Its prolific fall blooms make it very enticing to include in a garden, but be forewarned: it will quickly spread beyond the place where it was originally planted.

This species has rootlets with the same delicious taste, and inulin content, of its relative the Jerusalem artichoke. The seeds are a rich source of food for wildlife, including birds and deer, which also use the plant colonies for cover.

Heterotheca subaxillaris

CAMPHORWEED, TELEGRAPH PLANT, FALSE ARNICA

Camphorweed arrived in Texas from Central and South America, and it has spread across the southern United States. It is eclectic in its environmental preferences and does well in dry or moist conditions, especially in disturbed soil.

As its common name implies, camphorweed has a pleasant, pinelike, medicinal aroma derived from the various terpene compounds that it contains. The leaves are rough and slightly sticky (think of the texture of a cat tongue).

Camphorweed has anti-inflammatory properties. A tincture made with alcohol serves as a liniment for sprains, arthritis, and other such aches, or one can bathe in a strong tea for the same effect. A tea made from the whole plant can be used as a topical antiseptic to wash a cut or scrape. Drinking the tea purportedly relieves menstrual cramps as well as gas and diarrhea pains.

Camphorweed is a specific host to a charming brown and cream-colored herbivorous beetle, *Zygogramma heterothecae*, or camphorweed chrysomelid. The beetle is not affected by the anti-insect chemical

defenses of the plant, and each developmental stage of the beetle feeds on different parts of the plant.

A close-up of the stem and leaves is found on p. 112.

Isocoma pluriflora (*I. wrightii*)

JIMMYWEED, RAYLESS GOLDENROD

The colorful, mound-shaped jimmyweed is common in desert grasslands. Since it is not aggressive either in growth habit or seed production, it proliferates only in overgrazed pastures.

Jimmyweed contains a resinous material that is sticky to the touch and has a sharp turpentine smell. Both fresh and dry leaves contain a ketone poison that affects the nervous system of livestock. Animals that eat enough of it get the shakes and trembles, known as "the jimmies" in rural Texas. Affected animals also exhibit digestive upsets, difficulty in breathing, and dribbling of urine. The toxin passes through the milk to their offspring, and baby animals can exhibit symptoms well before the mother. Humans can be affected by drinking milk from cows or goats that have eaten jimmyweed.

Lactuca serriola

PRICKLY LETTUCE

Prickly lettuce is believed to be the ancestor of our salad lettuce, with which it will hybridize. This European invader species readily becomes a problem in cultivated areas.

The genus name *Lactuca*, which means "milk," refers to this plant's white sap, or latex, which has mild narcotic properties. The juice can be used as a sedative and as a cough suppressant. Pioneers also used the mildly acidic juice to treat warts, although it may cause dermatitis in people with sensitive skin.

The tender young leaves, which are high in vitamins A and C, may be eaten raw in a tossed salad. Older leaves are bitter and need to be boiled in several changes of water to make them palatable; seasoning with butter, salt, and bacon crumbles helps, too!

A view of the whole plant is found on p. 112.

Machaeranthera pinnatifida
YELLOW SPINY DAISY

The prickly character of yellow spiny daisy is best appreciated with magnification, and the origin of its scientific name, *Machaeranthera pinnatifida*, also becomes clear. Each tiny lobe of the deeply divided (pinnatifid) leaves ends in a tiny needle, reminiscent of a pawful of extended cat claws. Even the anthers (*anthera*) have sharp, swordlike appendages (*machaer*, meaning "dagger" or "razor").

Yellow spiny daisy is handsome in the spring when it first begins to bloom. Like its purple-flowered relative Tahoka daisy, it has the ability to continue blooming while forming seed heads from earlier flowers, so the plant becomes increasingly ratty looking as more and more of the flowers go to seed.

It is remarkably drought tolerant and blooms from spring through fall without regard to the moisture conditions. Yellow spiny daisy will perform well in a native plant landscape, but it seeds prolifically, so be prepared to thin out many seedlings in the next growing season.

Native Americans treated open wounds with a poultice made from crushed leaves. They also chewed the roots to relieve toothache pain.

A view of the whole plant is found on p. 112.

Psilostrophe tagetina

TEXAS PAPERFLOWER

Texas paperflower is an exquisitely pretty, tough, and tidy flower. It blooms even under very dry conditions and after a mild frost. Fine hairs covering the stem reduce moisture loss, giving it drought-tolerant properties, which in turn make it a delightful element in a xeriscape.

Paperflower makes an excellent dried-flower arrangement. To dry the entire plant, hang it upside down in a dark room or garage. The flowers retain their color well, and the petals keep their shape, becoming papery in texture, hence the common name.

Young plants contain enough sesquiterpene lactone toxins to make them poisonous to sheep, which become weak in the hindquarters, vomit, lapse into a coma, and eventually die.

A yellow dye can be made from the leaves and flowers.

A view of the whole plant is found on p. 113.

Pyrrhopappus pauciflorus (P. multicaulis)

TEXAS DANDELION, FALSE DANDELION

The Texas dandelion shares many characteristics with the common dandelion, but it is simple to tell them apart by comparing their anthers. Notice that the Texas dandelion has black anther tubes, and those of the dandelion are ordinary yellow.

The young leaves are edible and may be eaten raw in salads or boiled as a cooked green vegetable. The bitter taste of the older leaves can be remedied with two or three changes of water during boiling.

Rudbeckia hirta

RUDBECKIA, BROWN-EYED SUSAN

Rudbeckia is a native North American plant that made its way to Europe during the great age of exploration. Carolus Linnaeus named it in honor of his teacher and mentor, Olaus Rudbeck, who himself was a botanist and taught at the University of Uppsala in Sweden. Linnaeus served as tutor to the youngest of Rudbeck's twenty-four children in the 1730s.

Hirta means "hairy," and indeed, rudbeckia's leaves and stems are covered with hairs that have the texture of a cat's tongue.

Rudbeckia, Brown-eyed Susan

Brown-eyed Susan has attracted the attention of horticulturists since the 1830s in both Europe and America, and the commercial horticulture industry now offers a number of colorful cultivars. Its blossoms remain colorful for several weeks and lure many pollinators to the garden. It is simple to grow from seed, transplants readily, and thrives in poor soil with little water—all key attractions for flower lovers with non–green thumbs.

Rudbeckia has antibiotic properties against staphylococcal infections and may boost the body's immune response. It also has diuretic and cardiovascular stimulant properties.

The bright blooms of rudbeckia yield dyes ranging from tan to yellow or gray, depending on the mordant used to fix the color.

Senecio flaccidus (*S. douglasii* or *S. longilobus*)
THREADLEAF GROUNDSEL, DOUGLAS GROUNDSEL

The achenes, or "seeds," of the *Senecio* genus have fluffy white hairs attached. This trait mirrors the hair color of the geriatric generation and becomes the basis of the genus name *Senecio*, which is Latin for "old person."

Threadleaf groundsel may be found in overgrazed pastures, where it increases because livestock avoid eating these plants until all the more suitable forage is gone. In badly managed pastures or in disturbed sites, livestock may consume them and suffer severe liver poisoning. Cattle and horses are very sensitive to these toxins, but sheep and goats must eat a great deal more before they are affected.

Some *Senecio* species contain the alkaloid senecionine, which has a sedative effect. The Pueblo people made a paste of the leaves to treat sore muscles, irritated eyes, or upset stomachs, and they used parts of senecio stems to make arrows. Some species were believed to bring good luck in hunting and were smoked ceremonially.

This species performs well in xeriscape plantings.

Sonchus spp.

SOWTHISTLE, HARE'S THISTLE, HARE'S LETTUCE

Sowthistle, or hare's thistle, acquired its amusing common names from its attractiveness to pigs and rabbits, and indeed, in past centuries it was said to be able to restore mad (March) hares to their right mind. Its genus name, *Sonchus*, is based on the Greek word for "hollow," in reference to the drinking straw–like stems.

Sowthistle is a very common weed and enjoys a uniformly unpopular reputation among farmers throughout the world. Not only does it grow anywhere and everywhere but also the sap is especially beguiling to aphids, which flock to the flowers and exude their sweet and sticky excretions onto the surface of the flowering head. This stickiness contributes to the usual trashy appearance of sowthistle by catching every bit of dirt, fluff, and insect body parts that are wafting about on the breeze.

Sowthistle produces a milky sap containing various medicinal compounds, and people on every continent use this plant as a folk remedy for practically every disease known to humankind (that's only a slight exaggeration). In Tanzania the raw root is used to rid the body of parasitic worms, in China the sap is used as a treatment for opium addiction, in New Zealand it serves as chewing gum, and in sixteenth-century Europe the juice served as a beauty treatment. Then there are multitudes of the usual urinary, liver, heart, lung, snakebite, and snake oil recommendations common to many folk medicines.

Sowthistle, Hare's Thistle, Hare's Lettuce

Our garden vegetable lettuce is of the same tribal clan as this weed, so it should not be surprising that sowthistle's young, tender leaves can be recommended as a salad ingredient, as well as a good cooked vegetable, which is best when boiled in a couple of changes of water to remove any bitterness from the milky sap.

A yellow dye can be extracted from sowthistle.

A close-up of the mature seed head is found on p. 113.

Taraxacum officinale
DANDELION, LION'S TOOTH, WILD ENDIVE

Dandelion is a native of Europe and Asia but has been naturalized worldwide. The fact that Europeans valued this plant as an "official" medicine is reflected in its species name.

The leaves and roots are effective as a mild diuretic and can be used to treat problems of the urinary tract. The roots contain inulin, which serves as a dietary fiber that increases laxation and helps relieve constipation. Some believe the milky sap rubbed on warts and ringworm is an effective cure.

Dandelion is one of those uncharacteristic wild plants that can be consumed in large amounts without causing any apparent harm. Dandelion blossoms are delicious as fritters; battered, fried, and salted, and tasting like fried okra, they are sure to please any fast-food taste preferences. For a sweeter treat, roll the fried flowers in powdered sugar and dip them in syrup. Young leaves make a good green vegetable that is high in vitamins A and C as well as iron and calcium. For a beverage, roasted roots make a faux-coffee drink, and—taking the longer-term view—one might also make dandelion wine or schnapps.

Apparently taking into account the rapidity with which the yellow flower develops into a white puffball, a legend was created in which the dandelion represented a golden-haired maiden who in a short period of time turned into an old woman with white hair—the very parable of fleeting youth.

A pale yellow or tan dye may be obtained from this plant.

A view of the seeds preparing to take flight from the seed head is found on p. 113.

Tetraneuris scaposa var. *scaposa* (*Hymenoxys scaposa*)
PLAINS YELLOW DAISY

The genus name *Tetraneuris* refers to the "four nerves" or well-defined whitish or purple veins in the petals, making this genus both easy to identify and to remember. There are different forms of *T. scaposa* that are distinguished by the degree of definition of the veins on the underside of the petals.

The leaves of *Tetraneuris* have small glands that exude a bitter-tasting, unpleasant-smelling resin containing toxins that accumulate in the plant under drought conditions. However, sheep may develop a taste for this plant during the dry periods when their usual feed is not available, and they soon begin to show symptoms of poisoning. Its toxicity is not such a problem for humans. The outer stem and roots can be used as chewing gum, and a tea brewed from the plant can settle an upset stomach.

These daisies thrive in dry growing conditions and make a superb landscaping plant as a border along a sidewalk or at the front of a flower bed. As the flower ages, the ray flowers curl under or hang down, giving the maturing flowers a bedraggled appearance.

The flower heads yield a good-quality, lightfast, yellow dye.

The field appearance of plains yellow daisy is found on p. 113, with another variety (four-nerve daisy) showing prominent maroon-colored veins.

Thelesperma filifolium

GREENTHREAD

Greenthread is the bepetaled cousin of cota, mentioned in the next entry. The genus name *Thelesperma* is built from the Greek root *thele*, which means "nipple," and *sperma*, indicating seed, referring to the shape of the seeds. The species name *filifolium* refers to its threadlike leaves. The nodding buds are another agreeable and easily recognized characteristic.

The cheerful yellow flowers begin their blooming season early in spring and don't stop showing their faces until late summer.

The whole plant, including the flowers, yields dyes ranging in color from yellow to orange to a rusty brown.

Thelesperma megapotamicum

COTA, INDIAN TEA, RAYLESS GREENTHREAD

This slender, willowy plant is a cousin of greenthread, described in the previous entry, but unlike its relative, cota has small yellow flower heads that almost always have no ray flowers, or "petals."

Cota is most famous for the refreshing and flavorful tea that Native Americans in Arizona and New Mexico brew from the stems. The tea acts as a diuretic, and the Navajos use it for treating urinary disorders. Irritated skin may be soothed by soaking the afflicted part in cota tea. Drinking an especially strong brew will help to bring down a fever.

Cota yields rich, permanent orange, gold, and reddish-brown dyes that the Hopis, Navajos, and Jicarilla Apaches use to color cotton, wool, and basket fibers. Navajo wool weavers create various shades of orange by mixing cota with dock (*Rumex*) and mountain mahogany bark (*Cercocarpus*).

Tragopogon dubius

GOAT'S BEARD, JOSEPH'S FLOWER, YELLOW SALSIFY

Goat's beard is the exact translation of its scientific name (*trago*, meaning "goat," and *pogon*, meaning "beard"), referring to the long, white, parachute-like attachments on the seeds that look like the whiskers of a goat. The other common name, "Joseph's flower," refers to the belief that Joseph, husband of Mary, had a long, white beard resembling the puffball.

Goat's beard originated in Europe and has since become a naturalized citizen of North America. In Europe it was prized for its medicinal properties of aiding digestion and relieving heartburn, pleurisy, and liver disorders. The inulin in the roots provides the basis for a sweet-tasting syrup that is a good cough medicine for bronchitis. The roots also have diuretic properties for relieving water retention.

Native Americans used the sap as chewing gum. Young leaves and stems are good in salads or as a cooked vegetable, and the stem and buds can be eaten like asparagus. The young roots can be eaten raw, but the older roots need to be cooked. Pioneers, desperate for a coffee fix, roasted the roots and ground them as a coffee substitute, although they didn't get much of a caffeine buzz.

To get a good look at the large yellow blooms of the goat's beard, one must get out early in the morning, because it opens for business with the pollinating insects for only a few hours before closing shop as the day heats up.

A view of the large puffball is found on p. 113.

Verbesina encelioides

COWPEN DAISY, AÑIL DEL MUERTO, GOLDWEED

Cowpen daisy acquired its inelegant name from its habit of growing in disturbed areas such as livestock enclosures, oil fields, and roadsides, and from its faint "eau-de-bovine" perfume, which may cause allergic reactions.

Despite these unpleasant attributes, cowpen daisy has very useful medicinal properties. Native Americans used it as an anti-inflammatory to soothe insect and spider bites, boils, and mouth sores. Cowpen leaf tea may relieve gas and bloating, while skin ulcers and hemorrhoids can be treated with a poultice of boiled leaves. Cowpen daisy's phytochemical portfolio includes alkaloids with antimicrobial and anticancer properties.

Its robust growth habit and its propensity to seed prolifically ensure that it will crowd out less competitive neighbors. For easy landscaping of a dry, neglected alley, it's hard to beat.

A close-up of the stem and leaves is found on p. 113.

Xanthisma texana

SLEEPY DAISY, SUNNY DAISY

Sleepy daisy's genus name recognizes the loveliness of its lemon-colored flowers with the Greek root *xanth*, meaning "yellow." Its pejorative common name was bestowed because it "sleeps in" until about noon, when it finally opens its petals.

Xanthisma is a valuable addition to a landscape setting both for its lovely appearance and its ability to attract butterflies. When planted in bright sunny areas with sandy or caliche soil, where few other plants will thrive, sleepy daisies will shine. Don't try to get it to perform in shade; it will never wake up.

Zinnia grandiflora

PLAINS ZINNIA

Plains zinnia's genus name honors the German medical professor Johann G. Zinn, who made plant collections in Mexico in the 1700s. Plains zinnia blooms continuously throughout the year in the warmer parts of Texas and through the summer in northern Texas. It spreads by rhizomes and also reseeds itself, but it never becomes aggressively invasive.

This totally charming little plant is another landscaper's dream, especially for plantings in hot, dry environments. It forms short, compact, canary-colored bouquets that make outstanding borders or showy

Plains Zinnia

displays in mass plantings or rock gardens. When the flowers are spent, they remain daintily papery and dry on the plant. Butterflies are fond of plains zinnia as well.

A view of the whole plant is found on p. 114.

BERBERIDACEAE (Barberry Family)

Berberis (*Mahonia trifoliolata*)

AGARITA, ALGERITA, AGRITO

Agaritas are small evergreen shrubs armed with sharply pointed compound leaves reminiscent of holly. The yellow blooms give way to scarlet berries. The name "agrito" means "little sour," which gives a clue about the tartness of the fruits; these are prized for making wine, cobblers, pies, and preserves. Even the new leaf growth has a pleasantly sharp flavor and can complement a salad. Animals relish the sour red berries and help disperse the seeds in their droppings.

Agarita contains the highly bitter alkaloid berberine, which gives a bright yellow color to the root and wood. This chemical helps reduce fever and is an antibacterial and anti-inflammatory agent (especially for the gums) as well as a laxative. Agarita is a stimulant for the liver and a traditional remedy for the liver problems suffered by chronic heavy drinkers.

The Apaches decorated their skin for certain ceremonies with agarita's purple berry juice, and other Native Americans extracted a yellow dye to color wool for weaving. Small carvings are made from the exquisite yellow wood.

A view of the fruits and leaves is found on p. 114.

BIGNONIACEAE (Catalpa Family)

Tecoma stans

YELLOW BELLS, YELLOW TRUMPET FLOWER, ESPERANZA

Yellow bells is a deciduous shrub native to the tropics that has become naturalized in southern and far western Texas. It is now a favorite in landscapes and grows well in both high- and low-moisture conditions.

Bees and hummingbirds are fond of yellow bells. The leaves contain about 15 percent protein, and both leaves and flowers are nutritious browse for livestock.

All parts of the plant contain biologically active compounds that are used in traditional medicines. It is a common home remedy in Mexico for Type II insulin-resistant diabetes. Additionally, a tea made from the flowers and leaves helps to soothe indigestion brought on by heavy drinking. It is said to be effective against viral illnesses, such as colds and flu.

BORAGINACEAE (Forget-me-not Family)

Lithospermum incisum

PUCCOON, APACHE TEA

The puccoon's hard, nutlike seeds earned them the appellation "stone seed," or *Lithospermum*, and the name *incisum* refers to the deeply cut petal margins.

In the spring, the yellow, crinkled petals are showy, but the flowers are sterile; however, in the summer, the plant produces very small, inconspicuous flowers, which are fertile. Whether the flower was in the flamboyant stage or the more retiring mode, Native Americans would diligently look for it, not so much because of its food value, which was small, but because of its medicinal properties.

The root was brewed to make a tea and could also be boiled or roasted as a vegetable. People chewed the root to treat a cold and prepared an infusion of roots to alleviate stomach pains, lung diseases, or kidney problems. The roots and seeds were crushed and stirred into water to use as an eyewash. Native Americans apparently held the curative powers of puccoon in very high regard, as evidenced by their belief in its efficacy for restoring the use of paralyzed limbs.

Contrary to present public perception, birth control is not a modern invention. Apache women made a tea from puccoon to help space the arrival of their children. Hormonelike substances have been extracted from the plant's juices. One of the substances affects the ovaries, and another acts on the thyroid; both actions tend to inhibit pregnancy.

Native Americans also burned the dried plants as incense, fashioned beads from the hard seeds, and extracted a dark purple, blue, or red dye from the root.

CACTACEAE (Cactus Family)

Echinocactus wislizenii (Ferocactus wislizenii)
SOUTHWESTERN BARREL CACTUS

The southwestern barrel cactus is a stunning desert plant, with its massive, rounded body reaching up to two feet across. It may attain the ripe old age of one hundred, and as it matures, it becomes ever more cylindrical and tall, reaching six feet or more in height. This cactus makes its home in rocky soil, especially around El Paso and in Mexico and Arizona.

The flowers circle the top like a crown and range in color from yellow to bright orange or red. Following the convening of pollinators among the forest of anthers, the pineapple-shaped fruit develops and may remain on the cactus for a year—unless it becomes part of some animal's daily food intake. Javelinas, squirrels, birds, and deer are especially fond of the fruit.

The moisture-laden pulp may save someone from dying of thirst, but the slimy alkaline liquid also brings on diarrhea.

Opuntia phaeacantha

PRICKLY PEAR

Prickly pear and other members of the *Opuntia* genus are a nutritious food source for both humans and livestock, but the fare is well guarded by an armory of spines and prickles.

The stamens of the large *Opuntia* genus are thigmotropic: they are sensitive to pressure and move when touched. They may also be seen to move with no evidence of wind or other obvious cause, and a closer examination will reveal myriads of little insects burrowing deep into the flower, elbowing each other at the pollen feast.

The pads, or nopales, are available in many grocery stores, both fresh and pickled. The fresh nopales may be sliced and cooked in various ways. Their mucilaginous texture is reminiscent of that of boiled okra. The wine-colored fruits, or *tunas*, shown in the Exploring Further section on p. 114, have a sweetish flavor and are tasty eaten raw, roasted, or in jams. To avoid the glochids—those perniciously wicked little hairlike spines—skewer the fruit like a marshmallow and singe the prickles over

Opuntia phaeacantha, *prickly pear; red dye from crushed cochineal scale insects*

a fire; then peel and eat them or turn them into jam.

Opuntia has a fascinating host-guest relationship with the cochineal scale insect, a source of a rich red dye. The insects conceal themselves under white frothy fuzz that looks like bird droppings on the surface of the prickly pear and, when crushed, release a vibrant red dye that is as bright as blood, as the photo here depicts. The Aztecs and their neighbors scraped off this white excrescence and crushed the scale insects living within to obtain a scarlet pigment used to dye fabric. When the Spaniards arrived in the New World, they were intrigued by this rich, rare color and set their avaricious hearts on creating a trade monopoly with Europe. For two centuries cochineal dye was a valuable export second in importance only to gold. Today the dye is widely used as a natural colorant for food, beverages, and cosmetics throughout the world.

Prickly pear has a number of medicinal properties and has been used as a poultice for boils and wounds as well as a treatment for arthritis, eye problems, headaches, and sleeplessness. Bruises and burns can be effectively treated with fillet of cactus pad. The pads contain complex starch compounds that draw the fluid from the wound while soothing the skin. Putting a piece of peeled pad on mouth sores and infected gums works the same way.

Drinking prickly pear juice can relieve the pain of urinary infections or irritations, although it does not have an antibiotic effect. The juice also has a beneficial blood-sugar-lowering effect on obese, insulin-resistant individuals with adult-onset diabetes, and it also lowers LDL ("bad") cholesterol levels. Some Mexican Americans take advantage of these properties by regularly consuming diluted prickly pear juice, keeping it on hand in the refrigerator.

The gel inside the pads makes a good conditioner for the hair and is still used for this purpose by some Mexican Americans today.

Edible red fruits of prickly pear cactus are shown on p. 114, along with related *Opuntia* species, cowtongue prickly pear, and blind prickly pear.

Prickly pear cactus also has other flower colors that are pictured on p. 26.

Lesquerella fendleri

BLADDERPOD, POPWEED

Lesquerella's two common names, "bladderpod" and "popweed," refer to the little seedpods that form between the new flowers and the leaves. The pods look like inflated bladders and pop when they are stepped on.

As one of the first plants to bloom in the spring, bladderpod's fresh green foliage is avidly grazed by cattle. *Lesquerella* makes suitable forage except when consumed to excess; horses that overgraze this plant, for example, may develop swollen legs.

Although not really considered a food plant for humans, *Lesquerella's* seeds can add a peppery flavor to food.

Bladderpod seeds contain an oil that is rich in hydroxy fatty acids, which may prove to be economically important in the manufacture of plastics, coatings, lubricants, and cosmetics. In addition, the seed coat contains a gum that can be used as a thickening agent for food and for the recovery of crude oil.

CUCURBITACEAE (Gourd Family)

Cucurbita foetidissima
BUFFALO GOURD, CALABAZILLA

The leaves of buffalo gourd are foul smelling, a fact commemorated in its species name, *foetidissima*. Like many members of the gourd family, buffalo gourds are monoecious, with separate male and female flowers on the same plant. The female flower parts fall away from the ovary as it matures into a fruit.

Native Americans roasted and ate the nutty-flavored seeds, which contain about 30 percent protein. The seeds also contain a large amount of oil, which can be extracted for cooking as well as for hair care.

The large, carrot-shaped root can reach a whopping ninety pounds in weight and may contain as much as 20 percent carbohydrates. This monstrous root has been treated with mystical respect, and the Omaha, Ponca, and Dakota peoples made an offering of tobacco before harvesting it and took great care not to injure it during the extraction process.

Buffalo gourd leaves, stems, and roots all have laxative properties. For external use, the plant was crushed and applied to sores or ulcers in the skin. Rio Grande Valley Indians treated rheumatism by baking the fruits, splitting them, and placing them on the afflicted joints. As horrifying as it sounds, sometimes maggots would infest wounds, and an extract from boiled gourd roots was used to kill them. This same liquid was also used to treat earaches. If a person were unfortunate enough to become the host to parasitic worms, such as tapeworms and other such unwelcome intestinal passengers, he or she could eliminate the "guests" by consuming gourd seeds that had been crushed and put in water. A homemade remedy for treating sore hooves in horses was to have the animal stand with its foot in a bucket full of boiled gourd roots.

Native Americans used the roots, which contain a sudsy substance called saponin, for washing clothes and for shampooing the hair. The roots also contain starch, which was extracted for stiffening fabric and for thickening food and making puddings. Because of its high starch and oil content, buffalo gourd continues to be investigated as a potential new crop for dry areas.

The dried gourds have a tough outer skin and are useful in making cups, ladles, and other handy items. Native American dancers have also used them as leg rattles.

The cutaway view of the female flower, revealing the stigma and developing fruit, are pictured in the Exploring Further section, along with the baseball-sized mature fruit, on p. 114.

Ibervillea lindheimeri

BALSAM GOURD, YERBA DE VIBRONA, COYOTE MELON

Balsam gourd enchants wildflower aficionados, not with its stunning beauty, which it lacks altogether, but with the surprise of its little red fruits that appear on the vines in late summer and with the quaint charm of its squatty, bulbous stem partially buried in the ground. It grows in open areas or canyon breaks from Central and West Texas to the Panhandle, as well as in Mexico and the southwestern United States.

The plants are dioecious, with the males and females on separate plants. The female plants are easily recognized when in flower by the small gourd-shaped ovary beneath the flower. This green-striped ovary will mature into a hot orange or red fruit. A plant growing in favorable circumstances may be decorated with forty or more fruits.

Not so obvious is the stem, which is mostly underground and may be up to a foot in diameter in plants that are several years old. It is often called the "root," but it is actually the stem, from which the branches (vines) grow. The vitality of these stems is such that they will continue to send out branches year after year even if no water is available.

The stem of a closely related species, *I. sonorae*, also known as *huereque*, contains substances that help control Type II diabetes, and it is commonly used by Hispanics in Mexico and the southwestern United States for this purpose.

Balsam gourd has caught the attention of landscapers and cucurbit fanciers and is now marketed commercially for home gardens. It plays a small role as food for quail, which eat the seeds, and deer, which browse the leaves.

A view of the mature fruit is found on p. 115.

FABACEAE or Leguminosae (Bean or Pea Family)

Chamaecrista fasciculata (*Cassia fasciculata* or *C. chamaecrista*)
PARTRIDGE PEA, BEACH SENSITIVE PEA, GOLDEN CASSIA, SLEEPINGPLANT

The blooms of partridge pea have five yellow petals, with one being somewhat larger than the others and with one petal curled coquettishly over the stamens. Near the base of the petals is an eye-catching dark red mark. Rather than extend straight up, the yellow or purple stamens lean to one side. The pinnately compound leaves are sensitive to touch and will fold when brushed and also when the plant "sleeps" at night. This fascinating phenomenon has led to a few of its common names, such as "sleepingplant" and "beach sensitive pea."

Partridge pea inhabits the eastern half of the United States, growing in sandy, nutrient-poor soil. It is drought tolerant, grows to about thirty inches in height, and makes an excellent choice for a native plant land-scape with its attractive floral display through the summer and fall. A gardener should collect the seeds when the pods have turned brown but not yet split open. Partridge pea readily reseeds, so it is wise to deadhead the plants (remove the spent flowers) to prevent invasion into unintended

Partridge Pea, Beach Sensitive Pea, Golden Cassia, Sleepingplant

areas. The extensive root system has two positive ecological benefits: it is a good soil stabilizer, and it forms a symbiotic relationship with nitrogen-fixing bacteria, which enrich the soil.

The blooms are about an inch across and attract bees and butterflies. In fact, some butterflies actively seek out partridge pea as a host plant for their larvae. It is not unusual to find ants feeding at the glands near the leaf bases. The seeds provide food for many bird species, but the presence of anthraquinone causes gastric distress and diarrhea in mammals if eaten to excess.

Native Americans made a tea from the leaves to treat weakness, dizziness, and nausea.

Hoffmanseggia jamesii
JAMES' HOG POTATO, HOG POTATO

Hog potato, a small perennial growing from about six inches to one foot in height, is rather more attractive than its name would indicate. Its fern-like, bipinnately divided leaves are delicate looking, its reddish stem is a surprise, and the showy flowers with gold to orange petals adorned with red spots always call for a second look. The seedpods are dotted with glands, giving them a sticky texture. However, the most interesting feature of this little weed is what lies beneath the ground. A foot or so below the surface are small, potato-like tubers, which are relished by hogs— hence the common name. As one would expect, some Native Americans harvested the potatoes and roasted them as a carbohydrate source.

When one begins to notice this plant's range of habitat, it becomes evident that even harsh environments are no deterrent to its growth. Hog potatoes are even able to push up asphalt until it breaks, and then they spring up through the cracks. They are a lesson in perseverance under adversity!

A close-up of the flowers of a related species, *H. glauca*, is shown on p. 115.

Melilotus officinalis

YELLOW SWEET CLOVER

The Latin word for "honey," *mel*, is evident in the genus name; in fact, honey made from the nectar of *Melilotus* is considered among the best in the world. A free-seeding native of Europe and Asia, yellow sweet clover has become naturalized throughout North America. Another species, *M. albus*, has white flowers, but is otherwise very similar to *M. officinalis*.

Sweet clover's use in the manufacture of several food products, including cheese and alcoholic beverages, reflects its European origins. Leaves are used to make Gruyère and Schabzieger cheeses, and vodka steeped with this plant is very similar to a sweet traditional vodka made in Poland and neighboring Russia.

The smell of the plant, reminiscent of vanilla, is due to the presence of coumarin, a blood-thinning agent. Sweet clover has been used for treating inflamed and swollen joints, eye problems, earaches, and digestive upsets and as a poultice for sore breasts caused by mastitis.

Yellow sweet clover was an important forage crop in Europe as long ago as the sixteenth century and is still considered valuable for livestock

feed. Improved varieties of sweet clover that have less coumarin reduce the bloating problems that occur when cattle graze the wild types.

Early settlers stored dried yellow sweet clover with bed linens to give them a pleasant fragrance as well as to repel bedbugs.

Prosopis glandulosa

MESQUITE

Mesquite is a ubiquitous feature of the southern Central Plains and is found throughout Texas. It varies in size from a small shrub to a tree, depending on the availability of water.

One group of Native Americans, the Jumas, used mesquite wood for making cradles, and it is still used to make jewelry, small furniture, flooring, and paneling. Native Americans used the inner bark to provide fiber for making fabric and baskets. The robust spines were modified to serve as sewing needles.

American Indians ground the seeds to make a flour or meal, known as pinole, from which they made bread. The flowers may be eaten raw, and so can the beans, which are high in sugar. Delicious syrup can be made by boiling the pods in water until the liquid is condensed.

A high-quality hydrocolloid, or gum, similar in quality to gum arabic, can be produced from the sap. Native Americans used the gum as a source of black dye for fibers, as an adhesive for repairing broken pottery, and as a medicine for sore throat, stomach ulcers, diarrhea, and skin injuries. Pimas made use of the dark resin for a hair dye and as a wash for irritated eyes and skin. They applied powdered resin to the navel of a newborn to prevent infection.

The pods are a valuable livestock food. Mesquite seeds germinate much better after passing through the digestive tract of cattle, where they are scarified by the stomach acids; this is believed to have contributed to mesquite's spread through the cattle country of Texas.

The wood is prized for use in barbecues because it burns slowly and gives a smoky flavor to the meat. Mesquite also serves as a prognosticator of spring: ranchers know that when the old mesquites begin to put out leaves, the danger of frost is past.

A view of the edible ripe beans is shown on p. 115.

Senna roemeriana (Cassia roemeriana)

TWO-LEAFED SENNA

Two-leafed senna is named for its pairs of leaflets, which look like a "V for Victory" sign. The species name *roemeriana* commemorates the German naturalist Ferdinand Roemer, who collected plants in the New Braunfels area of Texas during the mid-1800s.

Two-leafed senna is toxic to cattle, affecting their muscles and digestive tract. Goats sometimes die of heart failure after feeding on this plant; however, horses and sheep are more apt to suffer liver damage. Despite its toxicity, a dilute tea made from the leaves has been used as a laxative.

This fetching plant is easy to grow, and its gold flowers make a colorful addition to a wildflower garden.

A view of the fruit (beans) is found on p. 115.

FUMARIACEAE (Bleeding Heart Family)

Corydalis aurea var. *occidentalis*

SCRAMBLED EGGS, GOLDEN SMOKE, GOLDEN CORYDALIS

Scrambled eggs are among the first plants to bloom in the spring, and their bright yellow flowers are a pleasure to see after a long winter. Its genus name, *Corydalis*, is derived from the Greek word for "crested lark," because the arched spur on the flower somewhat resembles a bird's feathery crest. The name *aurea* means "golden." As for its common name, "scrambled eggs" reflects its namesake both in color and general appearance.

Members of the genus *Corydalis* contain alkaloids that have narcotic properties. These toxic chemicals, found in the rhizomes, or underground stems, are used to treat inflammation and to relieve pain, including arthritic discomfort. *Corydalis* is occasionally used to calm nervous trembling, but self-administration is discouraged, as these toxins can be very dangerous. When Ojibwas suffered from some emotional upset, they would calm themselves by inhaling the smoke of the roots roasted over coals. Today, *curanderas* in Mexico administer *Corydalis* tea to women who have just given birth in order to help them regain their strength.

When sheep eat too much *Corydalis*, they may suffer from respiratory problems and convulsions, which may lead to death.

LINACEAE (Flax Family)

Linum hudsonioides
YELLOW SAND FLAX, TEXAS FLAX

Yellow sand flax and its cousin stiff-stem flax, shown in the Exploring Further pages, are both members of the same genus and share many traits, but they do have some distinguishing characteristics.

Both of these flax species may grow up to two feet tall. The cuplike flowers, borne in terminal clusters, are showy enough to gain them a respected place in a wildflower garden. They will do best in well-drained soil where there is plenty of sun. Yellow sand flax generally has fewer flowers per inflorescence than the somewhat more showy stiff-stem flax.

Stiff-stem flax contains an enzyme that releases cyanide, and when cattle, or especially sheep, graze it too intensively, they can be affected by the poison. Plants grown in warmer zones can also accumulate nitrates from the soil, which can result in oxygen depletion in the blood of grazing animals.

Both of these species are members of a useful genus that has given the world, both ancient and modern, the fibers from which linen is woven and the oil from which linseed oil is manufactured. Flaxseed and flaxseed oil have long been used in folk medicines to treat constipation, atherosclerosis (hardening of the arteries), high blood pressure, high cholesterol, kidney disease, menopause, and even cancer.

A close-up of stiff stem flax is found on p. 115.

LOASACEAE (Stickleaf Family)

Cevallia sinuata

STINGING CEVALLIA, STINGLEAF, STINGING SERPENT

Cevallia is a monotypic genus, meaning that it is represented by only one species. The mature seed heads of stinging cevallia are fascinatingly repugnant: the area where the seeds have detached resembles patches of sunburned skin.

Stinging cevallia grows in the same rocky limestone soil where its *Mentzelia* cousins can be found. The leaves of stinging cevallia have a sadistic unattractiveness: they give the appearance of thick, freshly cut cookie dough armed with wicked barbs.

The stiff hairs contain formic acid, and an intimate encounter with the hairs by an unprotected body part will result in the same sharp stinging sensation of being stung by ants. Application of a dilute solution of bleach is recommended for pain relief.

Mentzelia reverchonii

PRAIRIE STICKLEAF, REVERCHON'S STICKLEAF

Prairie stickleaf lives up to its name in two ways: the barbed hairs give the leaves the annoying ability to cling tenaciously to fabric or fur, and the mucilaginous sap may be used as an adhesive. This plant's common name is one that a person is not likely to forget—it sticks with you, so to speak.

Opening shortly before dusk, the gorgeous flowers of the prairie stickleaf unfurl their bright yellow petals to attract night-flying insects. The numerous stamens offer a generous supply of pollen, and after fertilization is accomplished, the seeds form in a seedpod that resembles a candle, complete with a wick from the remnants of the style. Stickleaf plants usually grow to about three feet in height.

Indigenous peoples of the Southwest harvested the oily seeds, which were generally eaten roasted but could also be consumed raw.

Prairie stickleaf was sought after as a treatment for sprains, arthritis, and aching joints. The entire plant could be made into a hot poultice for sore places. Some people just steep the whole plant in their hot bathwater and then steep themselves in the steaming water to get relief from general aches and pains.

Mentzelia oligosperma, a related species, is shown on p. 115.

ONAGRACEAE (Evening Primrose Family)

Calylophus berlandieri (C. drummondianus subsp. berlandieri)

SQUARE-BUD PRIMROSE, SUNDROPS, BERLANDIER'S SUNDROPS

Calylophus, the cousin of *Oenothera*, is named for the crested (*lophus*) calyx (*caly*) surrounding the unopened flower (the bud).

The shape of the stigma and the position of the stamens distinguish one genus of the evening primrose family from another. In the genus *Calylophus*, stamens are attached around the rim of the floral "cup," and the stigma is rounded, or "peltate," in contrast to the X-shaped stigma of the *Oenothera*. Species can be distinguished by the shape of the buds and the seedpods.

Like its *Oenothera* relatives, *Calylophus* flowers have extremely long floral tubes through which the style extends to the ovary far below. *Calylophus* shares other characteristics and properties with *Oenothera*. This information is provided in the next entry.

Photos of other evening primrose species are found on pp. 96 and 212.

Oenothera macrocarpa

RED-NECK EVENING PRIMROSE, COMANCHE CAMPFIRE, CANYON BLUE-LEAF PRIMROSE

The genus name *Oenothera* is based on the Greek word *oeno*, meaning "wine," which may indicate that this group of plants was used as a flavoring for wine or that eating them encouraged people to drink wine.

At first glance, the genus *Oenothera* looks very similar to *Calylophus*, but *Oenothera* can be distinguished by its X-shaped stigma.

Oenothera flowers have extremely long floral tubes, exemplified by the long "red neck" in the *O. macrocarpa* photo, through which the style extends downward to the distant ovary. Seedpods' shapes vary widely by species (e.g., quite large in *O.macrocarpa*, p.115).

The Navajos considered the evening primrose to be such an important healing plant that they called it a "Life Medicine." They used it to treat toothaches, irritated eyes, and sore throats, as well as digestive upsets and internal bleeding. Native Americans made hot poultices from the cooked plant to promote rapid healing of wounds, while boils were

treated with a primrose liniment. The boiled roots or leaves mixed with honey were used to quiet a cough. Evening primroses were also valued for their sedative, laxative, and diuretic benefits. Their efficacy as a medicine for heart disease, diabetes, premenstrual syndrome, and other diseases is related to the presence of gamma-linolenic acid in the oil, which is extracted from the seeds. Some evening primrose species are an economically important source for this nutraceutical oil.

The entire plant, including tender first-year taproots and young leaves, may be eaten raw or cooked with butter. The leaves add a peppery taste to salads. Native Americans and early settlers made use of evening primroses as nutritious green vegetables, and they drank a relaxing tea brewed from the leaves. The Hopis smoked leaves of this species as they would tobacco.

The seedpods and field appearance of *O. macrocarpa*, as well as the distinctive blue-gray leaves, are shown on pp. 115–116.

Another *Oenothera* species, *O. rhombipetala*, is shown on p. 116.

Photos of other evening primrose species are found on pp. 95 and 212.

OXALIDACEAE (Wood-Sorrel Family)

Oxalis corniculata var. *wrightii* (*O. dilenii*)
YELLOW WOOD-SORREL

Oxalis is taken from a Greek word meaning "sour," and its various Native American names all point to that same sour-salty taste profile. The word *sorrel* in its common name also means "sour" and is given to plants with a similar taste, such as *Rumex*.

All parts of yellow wood-sorrel contain oxalic acid and vitamin C (ascorbic acid), giving it a pleasant, tart taste reminiscent of lemons. This makes all parts of the plant a superb addition to a salad. *Oxalis* was a favorite treat of children of the Great Plains Indians, who ate any part of it, including the leaves and seedpods. On long treks the adults chewed the leaves to assuage thirst. Swellings were treated with poultices made from *Oxalis* flowers.

The oxalic acid is harmful to humans or livestock if consumed in excess, because it binds with calcium. However, a person would probably find consumption to be self-limiting, because the tongue swells and becomes numb if too much is eaten at one time.

Oxalis is easily propagated by creeping stems or tubers, and the species with larger leaves and flowers make attractive additions to a landscape. Ideal *Oxalis* habitat is a slightly shady area with loamy to clay soil and a steady supply of moisture throughout the year. The pH of constantly damp soil is usually low, and yellow wood-sorrel thus serves as an indicator plant for moist, acidic soil.

A view of the whole plant with seedpods is found on p. 116.

PAPAVERACEAE (Poppy Family)

Argemone mexicana

MEXICAN POPPY, YELLOW PRICKLY POPPY, CHICALOTE

The genus name comes from the Greek word *argemone*, which means "cataract of the eye," referring to the belief that the bitter yellow or orange sap could be used to treat eye problems. (Note: One should not attempt this treatment, since permanent damage may result.)

Showy, bright yellow blooms adorn the spiny, blue-green stems of the Mexican poppy, which grows in waste places and reaches over two feet in height. Even the buds are spiny and have three to five "horns" protruding from the top. It has become an international weed by spreading to Asia and Africa, where it is used for herbal remedies just as it is in its native territory.

When prickly poppy stems are broken, they exude a yellow sap containing a number of poisonous alkaloids and protein-digesting enzymes, which give the plant useful pharmaceutical properties. Native Americans used the sap to remove warts and to treat cold sores and other skin infections. A concoction of the flowers helps cold and flu sufferers rid them-

selves of the phlegm that is such an unpleasant accompaniment to these pulmonary complaints. The numerous tiny seeds can be used as a laxative, an emetic (substance used to induce vomiting), or a mild sedative.

A tea or wash with both analgesic and sedative properties is brewed from the entire plant. Those suffering from a bladder infection or prostate pain drink the tea to gain relief. Sunburned or scraped skin can be soothed by bathing the afflicted part with this liquid. The throbbing pain of a migraine sometimes responds to the tea. All these medical benefits led the Comanches to revere the plant, and they made offerings to it during harvesting.

The narcotic properties of poppies have led to many myths, one of which holds that the twin brothers Hypnos and Thanatos (Sleep and Death) of Greek mythology were crowned with this flower.

The only part of the plant suitable for food is the small, blackish-brown seed, which is generally roasted before being eaten. The seeds also serve as food for quail and doves. The plant is not a food source for grazing animals, however, due to the toxic components of the sap and the prickly nature of the stems and leaves. Since cattle avoid eating this plant, overgrazing may remove the competing flora, leading to the establishment of large colonies of poppies.

Native Americans reportedly used the spiny leaves dipped in ashes to make tattoos.

For photos of other colors of prickly poppy, see pp. 213 and 268.

PORTULACACEAE (Purslane Family)

Portulaca oleracea

COMMON PURSLANE

This prostrate succulent plant is well adapted to its role both as a weed and as the bane of gardeners. Common purslane forms thick mats that smother less aggressive plants, and it fails to wither and die as weeds normally do when subjected to the ministrations of a hoe or plow. Its prolific production of tiny black seeds, which can wait to germinate for a number of years, ensures its place on the roster of most obnoxious weeds.

Purslane is quite nutritious, with high levels of calcium, iron, and vitamins A and C. If one feels culinarily inclined toward purslane, there are a number of options, including boiling, steaming, frying, or pickling—or eating it fresh. The taste is tart, and its mucilage content makes a good thickener for soups. The ancient Pueblo people of the southwestern United States ate this succulent plant, boiling it with meat or stewing it as greens. Purslane is still sold as a fresh salad ingredient in markets in Mexico and the Caribbean. Because of its relatively high oxalic acid content, it must be eaten in moderation. Livestock deaths have been reported as a result of overconsumption.

The Navajos used purslane to treat stomachache and diarrhea. It was also used as a topical antiseptic.

RANUNCULACEAE (Buttercup Family)

Aquilegia hinckleyana

HINCKLEY COLUMBINE, YELLOW COLUMBINE, TEXAS GOLD COLUMBINE

Hinckley columbine's genus name, *Aquilegia*, is from the Latin word for "eagle," whose five long talons are mirrored in the spurs of the flower. This two-foot-tall native of the Big Bend area blooms early in the season to avoid the heat. Its intriguing flowers attract butterflies and hummingbirds and are bright spots in a landscape. In fact, it has become a commercially popular garden plant and has the added advantage of being deer resistant.

The Paiutes of the Four Corners region (where Arizona, Utah, New Mexico, and Colorado meet) used columbine as fragrance and as medicine. Sachets made from the aromatic seeds were stored with clothing. The Paiutes also chewed the seeds and rubbed them on their bodies and clothing to impart a pleasant scent. The roots of columbines were made into remedies for coughs and for rheumatic joints.

A view of the seedpods is found on p. 116.

RHAMNACEAE (Buckthorn Family)

Ziziphus obtusifolia

LOTEBUSH, GUMDROP TREE, CLEPE

The lotebush's long thorns and bluish-green leaves make this shrub easy to identify.

Native Americans dried the somewhat mealy textured fruit as a snack food. The seeds have a high oil content and may be added to breakfast cereals or breads. A number of wild birds include lotebush in their diet. Woodpeckers, in particular, search out lotebush fruits to feed their young. The thorns of lotebush provide a safe haven for nesting sites, with the added benefit of convenient food availability. Wood rats utilize the thorny branches for protective housing.

Soap may be obtained from the crushed roots, and the Apaches used it for shampoo. Root extracts also provide analgesic as well as anti-inflammatory medicines for treating wounds and skin ailments of both people and domestic animals.

Legend has it that the long-thorned lotebush was used to make the crown of thorns placed on Jesus' head at his crucifixion.

A view of the thorns and purple fruit is found on p. 116.

ROSACEAE (Rose Family)

Potentilla indica (Duchesnea indica)
INDIAN STRAWBERRY, FALSE STRAWBERRY, SNAKE BERRY

Looking very much like a "regular" strawberry (except that it has yellow flowers rather than white), Indian strawberry is apt to be quite a disappointment if the red fruit is popped into the mouth. Rather than the anticipated burst of flavorful sweetness, one will find that the fruit, while technically edible, is bland and insipid. The trifoliolate leaves, which also closely resemble those of the strawberry, may be cooked as a green vegetable.

As the scientific name suggests, members of the genus *Potentilla* are considered to be potent medicines. Skin infections and injuries have been treated with poultices and washes made from the plant. A tea brewed from Indian strawberry has been used for laryngitis, tonsillitis, and coughs. It has even been recommended for the treatment of snakebite, which explains why it is also called "snake berry."

This perennial makes an attractive ground cover, spreading rapidly by long stolons, or "runners." Indian strawberry does particularly well in shady, moist areas, but it has the propensity to become invasive.

A view of the strawberry-like fruit is found on p. 116.

SCROPHULARIACEAE (Snapdragon Family)

Verbascum thapsus

MULLEIN, FLANNEL MULLEIN, QUAKER'S ROUGE, BEGGAR'S BLANKET

Originating in Asia and Europe, mullein is now a common weed in North America. It spreads easily by multitudes of tiny seeds. The flowers are radially symmetrical, clearly an exception to the rule in family Scrophulariaceae. Mullein is a tall (up to eight feet) and compelling addition to a garden landscape.

More than two thousand years ago, Europeans recognized the value of mullein for medicine as well as other practical uses, and early North American settlers followed these traditions. The leaves have various bitter-tasting glycosides that are believed to have anti-inflammatory, antiallergenic, antioxidant, and antitumor effects. In order to treat asthma and other respiratory complaints, people smoked the leaves in pipes or burned and inhaled the fumes of the roots. The leaves and flowers can be boiled to make a tea for treating colds and bladder infections; however, a

tea made solely from the flowers packs more punch for cough relief. The flowers are also useful in relieving migraines. The root offers diuretic properties for eliminating excess fluid from the body.

Ancient Romans dipped the long flower stalks in tallow and used them as torches in funeral processions. Greeks used the leaves as candlewicks in oil lamps. Roman women extracted a yellow dye from the flowers to color their hair as well as fabric.

Mulleins have been put to more prosaic uses in recent times as well. Because Quaker women were not allowed to use makeup, they rubbed the fuzzy leaves on their cheeks to make them glow, so the plant was called "Quaker's rouge." Another common name, "beggar's blanket," is based on the use of the thick, furry leaves to line clothes to provide extra warmth. The soft, velvety leaves are sometimes known as "cowboy toilet paper," for obvious reasons. The leaves also make comfortable and renewable shoe liners, very similar to the way hummingbirds use the leaf hairs to cushion the lining of their nests. The tiny seeds of mullein are toxic and have been used (illegally) by lazy fishermen to stun fish.

A view of the whole plant is found on p. 116.

SOLANACEAE (Nightshade Family)

Nicotiana glauca
TREE TOBACCO, PUNCHE, COYOTE TOBACCO

Nicotiana glauca is a small tree that reaches about twenty feet in height. It is a native of Peru but has become naturalized in Texas and California. The leathery leaves contain the toxic alkaloid narcotic compounds nicotine and anabasine. Even the nectar of the tubular, greenish-yellow flowers contains the alkaloids, and the flowers have a heavy, musky scent.

Cattle and horses that make the mistake of ingesting the leaves suffer from trembling, frequent urination, vomiting, and diarrhea, or, in extreme cases, respiratory paralysis and death. Pregnant animals consuming tree tobacco may have offspring with birth defects.

In light of the toxic nature of tree tobacco, smoking it is clearly not recommended. However, there are those who have learned the art of fully fermenting and curing the leaves so they may be smoked without undue danger.

Native Americans found that crushed tree tobacco leaves could be placed between hot cloths and applied to joints and muscles for pain relief.

Tree tobacco also makes an excellent insecticide and has the added advantage that it does not carry the tobacco mosaic virus, which limits the use of regular tobacco as an insecticide for tomatoes, peppers, and other edible members of the nightshade family. Be warned, however, that spraying an extract of tree tobacco on your insect-ridden plants may stunt their growth.

Views of the whole plant and of mature fruits are shown on p. 117.

Physalis cinerascens
YELLOW GROUND CHERRY, POPWEED

The downward-facing little flowers of yellow ground cherry are easy to overlook, but it's worth the trouble to bend down and inspect both the flower and the fruit. When the ovary begins to mature and the petals fall away, the calyx enlarges and covers the developing berrylike fruit, which fills only half the enclosure. Some children enjoy popping this balloon-like structure, leading to the common name "popweed."

The ripe fruit is edible and can be made into jam, but be aware that the inflated calyx is toxic, so it should be peeled off and discarded. The

green unripe fruit as well as leaves and roots should also be avoided as potentially poisonous. The Pimas refused to eat the fruit, even when safely ripe, calling it "old man's testicles." The Great Lakes tribes were not so proud and used yellow ground cherry for food, and the Great Plains Indians ate the fruits as an appetite stimulant. They also burned the roots and inhaled the smoke for headaches and used the roots to dress wounds. The fruits have a diuretic effect and may be used to relieve water retention.

Solanum rostratum

BUFFALO BUR, KANSAS THISTLE

The bright yellow, star-shaped flower of buffalo bur is the only pretty thing about this weed, whose stems and leaves sprout a multitude of spines. The prickly seedpods, which are veritable porcupines, led to its common name, which is based on the belief that the burs became entangled in the fur of buffalo as they wallowed in their dust baths.

This drought-resistant plant is a serious weed in disturbed areas, and its toxic foliage and fruits can cause poisoning of livestock. In truth, such poisoning is rare, because grazing animals are rather put off by the presence of spines, but they may consume the early growth of this beastly plant. The presence of the glycoalkaloid solanine causes disturbances of both the nervous and the digestive systems, leading to trembling, excessive salivation, labored breathing, nausea, and in severe cases, progressive paralysis and death. Cattle are more sensitive to buffalo bur toxins than are sheep or goats.

Despite buffalo bur's toxic nature, the Zuñi Pueblo people brewed a tea from the powdered root to control nausea.

Buffalo bur is a host plant for Colorado potato beetle and is a problem in areas where potatoes are grown.

ZYGOPHYLLACEAE (Caltrop Family)

Larrea tridentata
CREOSOTE BUSH, CHAPARRAL, HEDIONDILLA, GOBERNADORA

Larrea is such an important member of the desert biome throughout the southwestern United States that one of its common names, "chaparral," has been bestowed on the ecological zone that it dominates at elevations of about three thousand feet. The desert severely rations every natural resource required by plants (except sunshine), and this tends to bring out the worst in plant behavior, as demonstrated in the way desert plants exhibit very unneighborly attitudes to fellow inhabitants in the scramble for food and water. *Larrea* is well equipped for biological warfare, and its allelopathic skills are such that the toxic oils that leach out into the soil around it restrict the germination and growth of almost every other plant species. In vast swaths of the desert, *Larrea* has become the sole proprietor, earning it the name gobernadora, or "governess," of the desert. A heavy rain that leaches the inhibitory *Larrea* compounds out of the soil is needed to allow seeds of other plant species to germinate.

The pungent aroma of the creosote bush has wonderful tones of "eau de medicine," although there are enough people who fail to appreciate it that it also has acquired the name hediondilla, or "little stinking plant."

Its high phenol content exhibits antitumor properties, especially for lung cancer. At the same time, the Centers for Disease Control have cited several cases of cancer resulting from prolonged use of *Larrea* as an herbal remedy. In fact, it was once used as a food additive for preserving freshness, but the FDA withdrew its GRAS (Generally Recognized as Safe) approval in 1970 because prolonged or regular consumption was shown to cause liver damage.

The leaves are heated in lard with beeswax to make an ointment for painful arthritic joints. The salve is also useful for ringworm, insect bites, and saddle or other skin sores. Creosote tea is used to treat colds and other pulmonary disorders.

Creosote is used as a wood preservative for utility poles and railroad ties, which if left untreated, will decay where they come in contact with the soil. Its protective effect is based on the antibiotic properties of its phenolic compounds, which inhibit the growth of the fungi and bacteria in the soil that break down cellulose in wood.

A desert view of the whole plant is found on p. 117.

Tribulus terrestris

GOATHEAD, PUNCTUREVINE

Goatheads are an introduced species that originated in the Old World deserts. The name "goathead" refers to the objectionable character of the seedpod, which features robust spines resembling the horns of a goat and which are capable of puncturing bare feet or bicycle tires. The scientific name, *Tribulus terrestris*, says it all: "troubler of the earth."

During periods of drought, goatheads accumulate nitrates that may lead to oxygen deprivation in grazing cattle, as the nitrates compete with

oxygen in binding to red blood cells. They may also cause liver damage, blindness, peeling of the skin, and the loss of lips and ears.

Despite the rather horrifying effects of consuming this plant, it has some surprising uses. *Tribulus terrestris* contains saponins, or toxic soaplike compounds, that have steroid effects on the human body. Its hormone-like properties are promoted on a commercial scale as a nutritional supplement used by bodybuilders. Goathead leaves and stems may be dried and powdered to make a mildly diuretic tea that is said to lower cholesterol as well as blood pressure.

A close-up of the well-armed seedpod is found on p. 117.

The default color for Texas flowers seems to be yellow—all shades of it. And they all seem to look alike, especially the ones in the sunflower family! This section provides some other clues to distinguishing one genus or species from another, using plant shape, leaf color and structure, and fruits.

Berlandiera lyrata, *chocolate daisy; mature flower head with black seeds beneath brown bracts, p. 49.*

Dyssodia pentachaeta, *parralena; note flower head atop a naked peduncle (stalk) with leaves on lower stem, p. 51.*

Engelmannia peristenia, *Engelmann daisy; field appearance showing petal-like rays that curl under at tips with age, p. 52.*

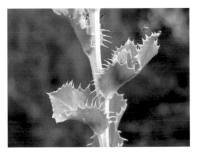

Grindelia ciliata, *sawleaf daisy; close-up of spine-edged leaves alternately arranged and clasping at the base, p. 53.*

Grindelia ciliata, *sawleaf daisy; field appearance of plants that may reach six to seven feet in height, p. 53.*

Gutierrezia *spp., broomweed; close-up of tiny flower clusters, p. 55.*

Helenium flexuosum, *purplehead sneezeweed; note down-flexed ray flowers, or "petals," p. 56.*

Helenium flexuosum, *purplehead sneezeweed; winged stems and alternate clasping leaves, p. 56.*

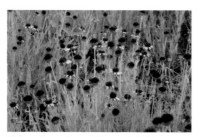

Helianthus ciliaris, *blueweed; field appearance showing grayish-blue leaves, p. 58.*

Helianthus ciliaris, *blueweed; mature seed head, p. 58.*

Heterotheca subaxillaris, *camphorweed; note glands on leaf surface, leaf bases clasping the stem, and flowering stalks arising from leaf axils, p. 60.*

Lactuca serriola, *prickly lettuce; field appearance of plants that grow up to six feet tall, p. 62.*

Machaeranthera pinnatifida, *yellow spiny daisy; field appearance of plants showing flowers in various stages of maturity, p. 63.*

Psilostrophe tagetina, *Texas paperflower; field appearance, p. 64.*

Sonchus *spp., sowthistle; mature flower head with fluffy pappus attached to seed-bearing achenes, p. 67.*

Taraxacum officinale, *dandelion; mature seed head showing parasol-like pappus attached to seed-bearing achenes, p. 69.*

Tetraneuris scaposa *var.* villosa, *four-nerve daisy; note prominent maroon veins on the underside of the petals, p. 70.*

Tetraneuris scaposa *var.* scaposa, *plains yellow daisy; field appearance showing dense basal leaves and bare flowering stalks with the petal-like rays on older flowering heads turning backward, p. 70.*

Tragopogon dubius, *goat's beard; note softball-sized "puffs" of mature seed heads and other stages of senescence, p. 72.*

Verbesina encelioides, *cowpen daisy; close-up of leaf arrangement, p. 74.*

Zinnia grandiflora, *plains zinnia; field appearance showing typically mounded plants, p. 75.*

Berberis trifoliolata, *agarita; fruits and three-parted spiny leaves, p. 76.*

Opuntia phaeacantha, *prickly pear; edible fruits, p. 80.*

Opuntia engelmannii *forma* linguiformis, *cowtongue prickly pear; novel shape and arrangement of the pads are food for the imagination—this one looks like a saguaro cactus! p. 80.*

Opuntia rufida, *blind prickly pear; areoles have glochids (barbed hairs) but not the long spines typical of cacti, p. 80.*

Cucurbita foetidissima, *buffalo gourd; female flower; note developing ovary, and the tendrils, used for climbing, p. 83.*

Cucurbita foetidissima, *buffalo gourd; cutaway view of female flower, p. 83.*

Cucurbita foetidissima, *buffalo gourd; mature fruit, p. 83.*

Ibervillea lindheimeri, *balsam gourd; mature fruit, p. 84.*

Hoffmanseggia glauca, *hog potato; note the profusion of red glands on the flowers, p. 87.*

Senna roemeriana, *two-leafed senna; dehiscent (splitting) seedpods and paired leaflets, p. 90.*

Prosopis glandulosa, *mesquite; edible ripe beans, p. 89.*

Linum rigidum, *stiff-stem flax; note shriveled petals on maturing seedpods, p. 92.*

Mentzelia oligosperma, *orange stickleaf; unlike other* Mentzelia *species, this one has only five petals and few anthers, p. 94.*

Oenothera macrocarpa, *red-neck evening primrose; large (macro) seedpods (carpa), p. 96.*

Oenothera macrocarpa, *red-neck evening primrose; field appearance, p. 96.*

Oenothera rhombipetala, *four-point evening primrose; note X-shaped stigma, long style, pointed petals, and the spotted sepals falling away to one side of the petals, p. 96.*

Oxalis corniculata, *yellow wood-sorrel; field appearance showing upright seedpods that resemble okra pods, p. 97.*

Aquilegia hinckleyana, *Hinckley columbine; dehiscent seedpods with filamentous styles still attached, p. 101.*

Ziziphus obtusifolia, *lotebush; note the fruit, which is a drupe with soft flesh around a single seed, and the leaves that are growing on robust thorns on branches, p. 102.*

Potentilla indica, *Indian strawberry; note fruit surrounded by leafy bracts and compound leaves with three leaflets, p. 103.*

Verbascum thapsus, *mullein; field appearance, p. 104.*

Nicotiana glauca, *tree tobacco; whole plant,*
p. 105.

Larrea tridentata, *creosote bush; field*
appearance showing typical uniform interplant
spacing brought about by internecine "chemical
warfare," p. 108.

Nicotiana glauca, *tree tobacco; seedpods,*
p. 105

Tribulus terrestris, *goathead; note seedpod*
(a sticker!) and pinnately compound leaves, all
covered with hairs, p. 109.

Green Flowers

AMARANTHACEAE (Pigweed Family)

Amaranthus spp.
PIGWEED, CARELESSWEED, ALEGRIA

Pigweed's genus name is much more flattering than its common one. *Amaranthus* comes from the Greek word meaning "unwithering," commemorating it as a symbol of eternal life. In fact, it was used in pagan burial rites and in magical spells to confer immortality. More prosaically, it is a highly favored food of pigs as well as cattle, sheep, goats—and humans.

The weedy species grow carelessly everywhere in disturbed soil and make a multitude of tiny black seeds, which can germinate even after twenty-five years. All species feature tiny green or red flowers growing on terminal spikes.

Amaranthus is also one of the oldest food plants in the New World, with evidence for its presence extending back two thousand years; during that time, Native Americans made genetic selections for seed-head size, leaf color, and different flavors.

The Pueblo Indians still consume both the seeds and leaves for their high protein content, as do the Incas of Peru. The Hopis and Navajos grind amaranth seeds to make a gruel, which is eaten with goat's milk, and they also parch the seeds to eat as a snack. Other Native Americans grind the seeds into flour for making cakes.

A tea from the leaves is an astringent, and it soothes stomach distress and relieves diarrhea. For babies with colic or vomiting, leaf tea in combination with lavender or feverfew flowers provides gentle relief. If the services of a poultice are required, the leaves will suffice—and provide a vegetable for the next meal.

Amaranthus accumulates nitrates, especially during drought. Ruminants, including cattle, sheep, and goats, are especially sensitive to nitrate poisoning, which results in oxygen depletion so that the animal in effect dies of suffocation. Symptoms manifest as unsteady movement, dilated pupils, and rapid pulse and in extreme cases, death, with the blood turning a chocolate-brown color. Pigweed also can cause abortion and bloat in ruminants.

Weedy *Amaranthus* is easily controlled with 2,4-D herbicide, but that chemical, too, will increase the nitrate level immediately after application.

ASTERACEAE (Sunflower Family)

Ambrosia grayi
BUR RAGWEED, WOOLLYLEAF BURSAGE

Bestowing the name "ambrosia," which means "food for the gods," on this plant was surely the deed of a cynic—until one remembers that the

mythological effect of consuming this heavenly food was to confer immortality. Therein lies the subtle clue in the name: ragweed is an invasive and noxious weed whose perniciously aggressive and perennial growth habit allows it to "live forever." It prolongs its generations both by seeds and underground stems (rhizomes) that produce dense stands able to crowd out annual crops.

Ragweed often establishes itself in disturbed habitats such as plowed fields or roadsides, as well as areas where water may accumulate, such as West Texas playa lakes or in moist clay soils throughout the Central Plains states.

Horses find this weed very palatable. It has been used as hay, particularly by those unable to afford more conventional forage, because in its early growth stages it does have nutritional value. Quail and other wild birds relish the seed as a winter food.

The juice extracted from ragweed has been used to stanch bleeding and to treat indigestion. Native Americans made a tea from the leaves and plant tips to reduce nausea, relieve constipation, and stimulate healing of skin sores. A powder from the root has analgesic properties and has been used for toothache. Plant extracts have astringent properties for use in treating diarrhea and other intestinal diseases.

The roots engage in chemical warfare (allelopathy) with neighboring plants by producing toxic substances that limit their growth. They particularly target those with nitrogen-fixing bacteria, as well as some grasses.

The male and female reproductive structures are found in separate flowers on the same plant but in different locations. In this monoecious arrangement, the males are located at the top of stems and often disappear by the time the females become noticeable with their multipronged, goatheadlike seeds. The male, or staminate, inflorescence is a prodigious producer of pollen that plagues those who suffer from hayfever.

The female, or pistillate, flowers produce small, spiny fruits that disperse themselves either by attaching to an unsuspecting accomplice or by floating on water. These seeds can survive a lengthy stay in the soil until favorable germination opportunities arise.

The seeds contain about 15 percent oil whose drying properties could be of potential use in varnishes and paints.

Artemisia filifolia

SAND SAGEBRUSH, SILVER SAGEBRUSH, WORMWOOD

Artemisia is named in honor of the Greek goddess Artemis, or Diana, the Hunter. The species name *filifolia* refers to the filamentous leaves.

A related species, *A. ludoviciana* (prairie sage, mugwort, St. John's plant), also has silvery foliage that makes it an attractive element in a xeriscape.

Undisturbed prairie areas, disturbed roadsides, and the home garden are all equally suitable habitats. *Artemisia* propagates prodigiously by rhizomes (underground stems) and can crowd out other species.

For centuries there has been a legendary association between *Artemisia* and John the Baptist, causing some to believe the plant had various magical properties and protective powers. In the Middle Ages, for example, it was believed that wearing a crown of its branches on St. John's Eve could prevent demon possession.

American Indians had the same beliefs in *Artemisia*'s protective capabilities over evil spirits and used it for many of the same medicinal purposes as did the Europeans. Like European women, Native American women used *Artemisia* for a number of gynecological problems. Mugwort is an old remedy for epilepsy, palsy, fever, upset stomach, and jaundice. It contains lactone glycosides that are antihelminthic and effective in expelling pinworms and other roundworms. *Artemisia* is also known for its antifungal and antibacterial properties.

Tea made from the leaves is quite bitter, but it once was a common substitute for regular tea among the European working class when the real thing became too expensive. *Artemisia* was used to flavor meats and was particularly prized as an addition to the stuffing mix used in baked geese.

Close-up views of both species of *Artemisia* are shown on p. 133.

Conyza canadensis

HORSEWEED, MULETAIL, BLOODSTANCH

What horseweed lacks in looks, it makes up for with a wide range of chemical constituents, some of which are useful pharmaceutical compounds and others of which are toxic to livestock. It makes itself at home on any soil type throughout all parts of the United States.

Young horseweed plants can cause significant losses among sheep, goats, and cattle during drought years, especially when the weeds have been sprayed with the herbicide 2,4-D. Brain damage is evidenced by extreme nervousness, walking in circles, and trembling and may result in coma and death. The B vitamin thiamine can relieve the symptoms if administered in time.

For humans, however, horseweed is useful for treating all manner of intestinal ailments, including diarrhea, hemorrhoids, diverticulitis, ulcerative colitis, and irritable bowel syndrome. It also is a remedy for meteorism—an old-fashioned euphemism for frequent passing of gas.

Horseweed has also been used to treat urinary tract problems in horses and humans, and its common name "bloodstanch" refers to its use in halting the excessive flow of blood. Some tribes of Native Americans used the leaves and flowers as smoking tobacco. The *U.S. Pharmacopoeia* in the 1800s listed the use of a horseweed oil extract for enhancing uterine contractions at childbirth.

Xanthium strumarium
COCKLEBUR, CADILLOS

Cocklebur's genus name, *Xanthium*, means "yellow," referring to the color of its sap, which was used by the ancient Greeks for hair dye.

The cocklebur is a robust cousin to the equally noxious weed *Ambrosia* (ragweed). Both have seed burs that are well adapted for hitchhiking: they easily become entangled in fur, or they can float on water to disperse their offspring far and wide. The mature burs may injure delicate tissues of animals, and their presence in sheep wool lowers its value.

Cocklebur plants house their male and female reproductive structures in separate flowers on the same plant, an arrangement described as "monoecious." The male flowers are at the top of the floral stalk, with female flowers growing in a cluster below.

The burs contain two seeds in a covering of tough, barbed spines; these two seeds germinate in successive years, making control of these weeds particularly frustrating. Seeds as well as seedlings contain a very toxic glycoside that inhibits germination and maintains dormancy in one

of the pair of seeds. Its concentration is highest in the cotyledons (first seed leaves), and the level drops rapidly in the seedling. The toxin disappears after the second true leaves appear.

If animals or humans consume cocklebur seeds or seedlings, the toxic glycoside may cause death from hyperglycemia (excess blood sugar). Unfortunately, cocklebur seedlings are very palatable to cattle, sheep, and swine, and death is not uncommon among those animals that browse the young plants.

The whole plant can be boiled and the liquid used to treat skin abrasions. An extract from the seedpods (burs) has been used as a diuretic to relieve water retention, as a treatment for diarrhea, and as a drink to soothe painful urination. However, cocklebur extracts may cause liver damage.

CHENOPODIACEAE (Goosefoot Family)

Atriplex canescens

FOUR-WING SALTBUSH, CHAMISO

Four-wing saltbush is named for the four "wings" attached to the small (1/4 to 1/2 inch) seedpod and for the fact that it grows well in saline or alkaline conditions. It extracts salt from the soil and deposits it on the leaf surface in little pouches that give this shrub its silver-gray color. Four-wing saltbush is dioecious, which means the male and female flowers are found on separate plants. The plant shown here is a heavily fruited female.

Four-wing saltbush was highly prized by the ancient peoples who inhabited what is now known as the Four Corners area of the United States, where Colorado, New Mexico, Utah, and Arizona meet. The seeds of this shrub have been found in many Pueblo sites, in the feces of the people who once lived there. The Hopis still eat its young leaves as greens and boil them with meat as a flavoring. Burned saltbush is the source of the ashes that serve as a substitute for baking soda. The Hopis also use the ashes to maintain the coloring in the blue-green cornmeal used for making piki bread. The Paiutes ground the saltbush seeds into a flour to make mush.

Navajos made a tea from the tips of young branches to relieve nausea and to break fever. They also made poultices to soothe insect bites and used the plant to treat stomach pains, toothaches, and coughs.

Pueblo people made a yellow dye from saltbush leaves and twigs.

The Zuñis had a superstition that rabbits, one of their meat sources, would multiply if prayer feathers were tied to the branches of four-wing saltbush.

Because it selectively takes up selenium from soils containing that element, saltbush may be used to reclaim land that has been previously unsuitable for agriculture. High selenium levels can therefore make it toxic, and sheep and cattle that use it for forage may suffer from bloating.

Close-ups of the male and female flowers are found on p. 134.

Chenopodium album

LAMBSQUARTERS

Some of Texas' wild plants are just plain ol' weeds that are despised by farmers and gardeners alike, and lambsquarters fits that description perfectly. But don't let its humble appearance fool you. This is actually a quite useful plant that is appreciated in many parts of the world for both food and medicine.

One of the authors lived for many years in Central Africa, where the women weed lambsquarters from fields and save the plants to cook for their families. The local name translates to mean "the husband is sleeping," a dig at the way the men leave the women to do most of the work to provide for the family. People here in Texas also learned long ago that lambsquarters is a very tasty green vegetable. The young leaves of this plant are delicious cooked with a little onion and some tomatoes. In addition to eating the leaves, Native Americans also gathered the seeds and made them into cakes and porridge.

Old-timers made a poultice from lambsquarters leaves to reduce swelling and pain from toothache. They also made a tea from the leaves

Lambsquarters

to relieve the pain of rheumatism or added the plant to hot bathwater as a soothing soak for the aches and pains of aging.

Old leaves can be consumed to cure stomachache, and eating seeds will help expel parasitic worms.

EPHEDRACEAE (Ephedra Family)

Ephedra aspera

ROUGH JOINT-FIR, MORMON TEA

Ephedra is more closely related to junipers, pines, cedars, and other gymnosperms than to flowering plants, since its reproductive structures are tiny cones rather than true flowers. *Ephedra* species can be distinguished by the shapes of these cones. *Ephedra antisyphilitica,* which has a fleshier cone than *E. aspera,* is shown on the Exploring Further pages at the end of this section.

The scalelike leaves are reduced to a mere two millimeters in length, so the plant carries on photosynthesis in its many slender, jointed green branches. These small shrubs are generally dioecious, with male and female cones occupying separate plants.

As the common name implies, a stimulating tea can be made from the green stems. This refreshing drink has also been used to treat urinary problems because of its diuretic and antiseptic properties, which has given rise to one of its common names: "kidney weed." *Ephedra antisyphilitica*, or "clapweed," received its "social" name because it was used to treat sexually transmitted diseases (STDs) in the days before antibiotics. It seems more likely, however, that it simply soothed the urinary tract pain that often accompanies STDs. In Mexico, *curanderas*, or folk herbalists, also recommend the tea for liver disorders.

The species of *Ephedra* found in Texas contain minute amounts of the alkaloid pseudoephedrine, an amphetamine-like drug that acts as an appetite suppressant. Its use in the weight loss agent "herbal fen-phen" has been banned by the FDA because of the danger of stroke, heart attack, seizures, and death associated with its use. However, pseudoephedrine's action as a decongestant for opening breathing passages and constricting blood vessels during a bout with a cold or flu is well recognized.

Native Americans roasted the seeds of mormon tea and ground them for flour. They also used the stems to make a pale tan dye.

A whole-plant view of *E. antisyphilitica* and its cones are found on p. 134.

EUPHORBIACEAE (Spurge Family)

Croton spp.
CROTON, SKUNKWEED, BARBASCO

Croton is the sort of weed that blends into the background and generally goes unnoticed. One would never guess the powerful wallop this plant can pack. Its oil delivers a very strong and dangerous laxative effect, and just a drop will produce an extreme purgative response. Croton oil may cause severe dermatitis when it comes in contact with the skin, but when diluted with olive oil or Vaseline, it may be used as a liniment or as a counter-irritant for arthritis pain. It may also act as an insect repellent: bedbug-infested mattresses or clothing can be cleared of their unwanted fauna by putting croton plants in their midst, or they can be fumigated with smoke from burning plants.

Croton oil was once used extensively as a medicine in Europe but is now considered so dangerous for human use that regulations are in place to prevent the oil from entering the food supply. For example, crotonic acid is used in the manufacture of polyvinyl chloride (PVC) plastic, but because crotonic acid is known to migrate out of the plastic into food, both U.S. and European food laws restrict the levels that are permitted in food products.

Both the Great Plains and the Pueblo Indians used the leaves to make a tea for stomachache and for bathing sick babies. The Pueblo people also used it for a purgative and for treating urinary problems and snakebite. The Zuñis devised a treatment for the post-European-contact disease syphilis: a person afflicted with that disease would drink a strong tea made from croton and the thistle *Cirsium ochrocentrum*, run until the point of sweating, and then wrap up in blankets to maintain an elevated body temperature for several hours. Subsequent research has shown that *Treponema*, the organism that causes syphilis, is killed when body temperature is sustained at 104–105°F for two hours.

Grazing animals avoid eating the toxic croton leaves even in the absence of other food. When the land is abused by overgrazing and croton's competing plants have been consumed, croton will replace the desirable vegetation. However, the seeds serve as food for doves and other wild birds that are not fazed by croton's toxic effects.

Close-ups of the male and female flowers are found on p. 134.

POLYGONACEAE (Buckwheat Family)

Rumex crispus

CURLY DOCK

Dock is native to Europe and Asia and has become naturalized in North America, including Texas. This two- to three-foot-tall plant is usually

found in ditches or other areas where water collects and is often surrounded by vegetation that obscures it until the seeds begin to ripen, and the formerly green inflorescence becomes a rich red-brown.

All parts of the plant are used for medicinal purposes. Oxalic acid is a principal active ingredient, along with tannins and vitamin C. The root has a gentle laxative action, while the soaked seeds may be eaten to stop diarrhea. Dock's astringent properties give it the ability to constrict blood vessels and make it useful for treating sore throats, coughs, colds, skin irritations, and gum inflammation. Some people have found relief for the pain of sunburn by treating the skin with grated dock roots.

Dock leaves can be eaten as a green vegetable but should always be cooked to reduce the sour-tasting oxalic acid content, which can be dangerous in excess. When grazed extensively by cattle, dock has been reported to cause kidney damage.

The roots are also a source of brown dye, and they have been used in tanning leather because of their high tannin content.

A field view of a related species, *R. hymenosepalus,* is found on p. 134. Note mature seed head.

TYPHACEAE (Cattail Family)

Typha latifolia

CATTAIL

Cattail's roots are adapted for living in ponds or poorly drained areas with standing water. Dense cattail colonies provide an important habitat for wildlife.

Young leaf buds of cattails may be boiled to provide a cabbagelike vegetable, and the green flower spikes taste something like corn on the cob. The pollen, which is high in protein, may be added to flour, and the

seeds are also a superior protein source. The underground stems, or rhizomes, are excellent sources of carbohydrates and may be cooked like a potato or ground into flour. However, one should never eat the rhizomes raw, as they may cause vomiting.

Native Americans treated burns with a mixture of fat and the crushed rhizome, which has a waxy coating. The long, slender leaves have been used in basket weaving. The seed fluff is used as tinder in starting campfires, and it has served as insulation and as the filler in life jackets, sleeping bags, and pillows.

Apaches consider the pollen to be sacred and use it in girls' hair in their coming-of-age ceremonies.

★ EXPLORING FURTHER ★

These plants were hiding behind the door when the large, colorful flowers were handed out. Maybe that's why so many in this group have such weedy personalities, from pigweed to cocklebur to ragweed! But even if they are not beauty contestants, many have valuable chemical properties. These close-ups of the generally tiny flowers will help you gain a greater understanding of these species.

Artemisia ludoviciana, *prairie sage; note the mitten-shaped leaves, p. 123.*

Artemisia filifolia, *sand sagebrush; note the narrow linear leaves, p. 123.*

Atriplex canescens, *four-wing saltbush; close-up of staminate flowers on a male plant,* p. 126.

Atriplex canescens, *four-wing saltbush; close-up of mature pistillate flowers on female plant showing the four wings on the fruit,* p. 126.

Ephedra antisyphilitica, *mormon tea; whole-plant view showing that every day is a bad hair day!* p. 129.

Ephedra antisyphilitica, *mormon tea; fleshy orange female cone,* p. 129.

Croton *spp., croton; note swollen ovaries of female flowers,* p. 130.

Croton *spp., croton; note cream-colored stamens on male flowers,* p. 130.

Rumex hymenosepalus, *canaigre; mature seed head,* p. 131.

Blue Flowers

ASTERACEAE (Sunflower Family)

Centaurea cyanus

CORNFLOWER, BACHELOR BUTTON, STAR THISTLE

Cornflower's genus name, *Centaurea*, is based on the mythological Greek centaur Chiron, half man and half horse, who was believed to be the father of medicine and veterinary science.

Cornflower came to the United States from the Mediterranean region and Europe, where, despite the beautiful blue flowers, it was despised as a weed in past centuries. In fact, it was called "hurtsickle" because the wiry stems blunted farmers' sickles during the grain harvest. It quickly became naturalized in this country and has since redeemed itself not only as a wonderful roadside flower but also as a domesticated addition to home gardens.

The leaves, seeds, and flowers are used for a variety of herbal remedies. Several pharmaceutically valuable chemicals, including phytosterols (for lowering cholesterol) and coumarin (a blood thinner) are among the compounds in its medicine chest. Cornflower stimulates the appetite and relieves indigestion. It also contains compounds with antibiotic properties. Cornflower leaves can be used to treat skin wounds: the high tannin content of crushed leaves gives them astringent properties that halt bleeding by causing tissue contraction. There are several reports of the plant's value as an eyewash or eye treatment.

Cornflowers retain their blue color when dried, especially when pretreated with glycerine. A blue dye can be made from the flowers, but it is not colorfast.

Cichorium intybus

CHICORY

Chicory joined the early wave of immigrants to North America from its home in Europe and has so adapted itself across the continent that it seems like a native. Chicory has also been cultivated in Egypt for thousands of years, and both the genus and species names are of Arabic origin.

Chicory has milky sap, a characteristic that it shares with its relatives, including lettuce, dandelion, goat's beard, and sowthistle. The blooms open before dawn and close by early afternoon except on cool, cloudy days, when they may remain open longer.

Members of the chicory tribe contain inulin, a fructose-based type of starch, but only this species stores enough in the root to be of commercial interest. Diabetics benefit from inulin-based products because they help maintain proper blood-sugar levels, and people who are prone to osteoporosis can include inulin or its derivative, fructo-oligosaccharide (FOS, for easier pronunciation), in their diet to improve calcium absorption. These products are commercially available in the United States thanks to imports from Europe, where several multinational companies also extract inulin from chicory root to make fructose syrups, which are sweeter than ordinary sugar. In addition, inulin is a source of indigestible fiber that makes a great laxative.

Chicory has a long tradition of use in salads and as a substitute for coffee, as well as a sedative and a diuretic. A yellow dye can be extracted from chicory when it is in full bloom.

COMMELINACEAE (Dayflower Family)

Commelina erecta var. *angustifolia*
ERECT DAYFLOWER, HIERBA DEL POLLO, WIDOW'S TEARS

In the 1700s when taxonomist Carolus Linnaeus was exercising his skills in the art of naming plants, one of his favorite things to do was to create Latin names commemorating people whom he knew and admired. In 1735 he became familiar with the plant collections of Johannes Commelijn and his nephew Caspar, the famous and highly respected curators of the botanical gardens in Amsterdam from the mid-1600s to the mid-1700s. Linnaeus classified a group of dayflowers as *Commelina* in their honor, with each of the flower's two large petals representing Johannes and Caspar. The third and tiny petal served to commemorate one of their relatives who died young, having made no significant contribution to the world of botany.

These beautiful flowers are one of the "here today and gone tomorrow" types, giving rise to the name "dayflower." Fortunately, the plant produces a steady supply of blooms all through the spring and summer. In the clasping modified leaf, or spathe, that encloses the flower, a sticky liquid accumulates; when squeezed gently, the spathe will release a drop of water—a "widow's tear." Dayflowers love moist soil, especially at the edges of shady areas. It makes a nice "filler" plant and will reseed easily.

The leaves, as well as stems and flower parts, may be chopped and added raw or cooked to vegetable dishes. The plant is rather mucilaginous, but if one can abide the slime of boiled okra, this property won't inhibit the appetite. The roots contain saponins (soapy alkaloids) and are best avoided as a food source. Dayflower's use as a folk medicine is limited, but there are reports of antibacterial activity from extracts of the flowers as well as of its use in lowering blood pressure.

Commelina erecta, *erect dayflower; whole-plant view*

Tinantia anomala, *annual widow's tears; note types of anthers*

In the field appearance photo of *C. erecta* (erect dayflower) shown above, note the parallel leaf venation typical of monocots and the boat-like spathe under the flower. Enjoy the cute fuzzy face of dayflower's cousin, *Tinantia anomala* (annual widow's tears), depicted beside it. What one is actually seeing are three bearded stamens. The other three stamens are clean shaven.

CONVOLVULACEAE (Morning-Glory Family)

Ipomoea violacea (*I. tricolor*)

HEAVENLY BLUE MORNING-GLORY, GRANNYVINE, BADOH NEGRO

The many species of morning-glory are found in temperate and tropical regions around the world. *Ipomoea violacea*, in particular, is a native of Mexico and has become naturalized from Texas to Florida.

The hue of the blue morning-glory varies mainly as a function of the age of the bloom. When the soft-swirl-shaped corolla first unfurls in the morning, it is a lovely blue, but as the hours go by, a pinkish undertone starts to show through, and an older flower will be more lavender. The center of the corolla is white. With gorgeous flowers and petal colors ranging from white to pink, blue, or purple—or even multicolored—morning-glory is a popular ornamental vine that self-seeds and readily escapes cultivation. It is also well appreciated by bees, butterflies, and birds. It grows very quickly and can easily climb fifteen feet and cover a fence or other plants in only a few days. With these qualities, it can be invasive and is a troublesome weed in cultivated areas.

Members of this species contain potent hallucinogenic alkaloids in their seeds. Indigenous peoples of Mexico and South America took advantage of this property to attempt to communicate with their deities in their religious rituals. Morning-glory also acts as an intestinal purgative.

The morning-glory is thought by the Japanese to symbolize immortality, while in English tradition it represents the stages of human life in a single day: a bud in the morning, a flower at noon, and wilted by evening.

For photos of other members of *Ipomoea*, see pp. 196 and 249.

IRIDACEAE (Iris Family)

Sisyrinchium spp.
BLUE-EYED GRASS

The name "blue-eyed *grass*" is deceptive because it gives the impression that this plant is a true grass. While it does have grasslike leaves, it is a member of the iris family and has beautiful petals, which no grass has. The genus name *Sisyrinchium* means "pig snout" in Greek, an ironic appellation for so pretty a flower. Pigs find this plant's attractiveness underground and relish the plant's yellow roots.

The many species of this genus hybridize readily, so it is not always easy to distinguish one species from another. It blooms in the months of March and April, growing in small colonies and self-sowing its seeds, ensuring a prolific return the following year.

A close-up of the flower is found on p. 147.

FABACEAE or Leguminosae (Bean or Pea Family)

Lupinus texensis

TEXAS BLUEBONNET, BUFFALO CLOVER, WOLF FLOWER

Bluebonnets became the Texas state flower in 1901. There are five species of genus *Lupinus* making Texas their home, and in 1971 state lawmakers decided to embrace them all as the state flower(s). Two of these are shown on the next page.

Despite the name, bluebonnets are not always blue; some species or varieties have pink, maroon, or white petals. Each new flower has a white spot, and careful observation will reveal that on some of the flowers this white spot has turned red, indicating that it has already been pollinated and is on its way to making a seedpod. All lupines have palmately compound leaves.

To grow bluebonnets in a home garden, plant the seeds in the fall. They will germinate with fall rains and grow through the winter as a rosette, or cluster of leaves hugging the ground. When the days lengthen, so will the flowering stalk, and the plant will begin to flower sometime

Lupinus concinnus, *annual lupine*   Lupinus havardii, *Big Bend bluebonnet*

in March or April. Bluebonnets will reseed themselves thereafter if conditions are favorable.

Many lupines contain dangerous alkaloids, but bluebonnets are "sweet"—that is, their alkaloid content is low—making them a desirable forage plant.

The Navajo people believed that lupines were a cure for sterility and would help a man produce female children. They also ground the roots and applied the paste to skin wounds. Wool may be dyed a light yellowish-green by simmering it with bluebonnet flowers, although colorfastness is a problem.

Legends abound about how the lovely bluebonnet came into existence. One story explains that the flowers are chunks of sky knocked down by warriors fighting in the Happy Hunting Ground. Another legend is about an Aztec maiden who was being sacrificed to appease the gods. She dropped her blue headdress to the ground, and the next day there were blue flowers growing there. A third well-known myth says the flowers arose from the ashes of a doll sacrificed by a little girl to bring the rains back to her tribal lands.

Two other bluebonnet species include *L. concinnus*, annual lupine, which is the least flamboyant of the lupines, and *L. havardii*, Big Bend bluebonnet, which is larger and more robust than other species. Both of these species are shown above.

Occasionally one sees white or pink bluebonnets, which show recessive genes for alternative colors. Examples of these are shown on p. 147, along with typical palmately compound leaves and seedpods.

GENTIANACEAE (Gentian Family)

Eustoma exaltatum

BLUEBELL, PRAIRIE GENTIAN, LIRA DE SAN PEDRO

The bluebell's lovely, large blossoms adorn the Central Texas countryside from June to September where moist soil is to be found. The flowers are usually bluish-purple, but they also come in lighter colors, even white and yellow. Commercial growers cultivate bluebells for the ornamental garden–plant market as well as for the cut-flower market. In Japan they are especially prized in flower arrangements.

Bluebells are the inspiration for the name of the famous Blue Bell ice cream manufactured in Brenham.

LAMIACEAE (Mint Family)

Salvia farinacea
MEALY SAGE

Mealy sage occurs throughout most of Texas, particularly in areas with limestone soil. Several varieties of mealy sage are now commercially available in a range of blues and purples, and with plants reaching two to three feet in height, they look very handsome in massed plantings. These mint family members do well in full sun or partial shade and require well-drained soil.

The fresh leaves contain vitamins A and C and can make a healthy addition to a salad or may also be used to flavor meats. As with all types of sage, the fragrant leaves make a refreshing tea. Sage tea also has antimicrobial properties and may be gargled to treat a sore throat or drunk to treat upset stomach and diarrhea. Nursing mothers have used the tea to help decrease breast milk during weaning.

Salvia texana

TEXAS SAGE, BLUE SAGE

Texas sage makes its home in the rocky soils of the South Texas Plains and the Edwards Plateau, growing to about a foot in height and forming dense clumps of somewhat branching stems. Its covering of relatively long hairs surrounds the green stems and leaves with a haze of white. Texas sage has the square stem and whorled leaf arrangement typical of the mint family, and, like those of its relatives, its flowers are strongly bilateral in symmetry.

LINACEAE (Flax Family)

Linum pratense

BLUE FLAX, MEADOW FLAX

Blue flax's genus name, *Linum*, is the Latin name for "flax" and is the root word for "linen" and "linseed oil," both of which are products of this family. The name *pratense* means "growing or found in meadows," a description that is reiterated in its common name, "meadow flax." This plant is apt to be found in dry, open places such as prairies, pastures, and roadsides.

Traditional healers grind flax seeds into a paste for plastering on wounds, boils, and any inflamed skin.

POLEMONIACEAE (Phlox Family)

Gilia rigidula

BLUE GILIA

The genus *Gilia* was named for a Spanish botanist, Philipp Salvador Gil, who lived during the 1700s. The western half of Texas is home to blue gilia, whose low-growing mounds of numerous star-shaped blooms brighten the roadsides, in contrast to the rocky or caliche soil where they often grow. This perennial usually grows low to the ground but may reach to a foot or more in height. Each five-petaled blue flower has an "eye" in the center, typical of the phlox family. Blooming in the spring and sometimes again in the fall, blue gilia makes a colorful addition to a rock garden.

A view of the whole plant is found on p. 147.

These photos of whole-plant views, leaf shapes, and alternative color presentations will help with identification in the field. *Commelina* and *Sisyrinchium* are monocots and have parallel venation, in contrast to the netted venation of dicots.

Sisyrinchium *spp., blue-eyed grass; close-up of the flower, p. 140.*

Lupinus harvardii, *Big Bend bluebonnet; pink phenotype, p. 141.*

Lupinus havardii, *Big Bend bluebonnet; white phenotype, p. 141.*

Lupinus texensis, *Texas bluebonnet; palmately compound leaves, p. 141.*

Lupinus texensis, *Texas bluebonnet; hairy seedpods, p. 141.*

Gilia rigidula, *blue gilia; field appearance showing narrow leaves, p. 146.*

Purple Flowers

ASCLEPIADACEAE (Milkweed Family)

Asclepias brachystephana

PURPLE-FLOWERED MILKWEED, LONE STAR MILKWEED, BRACT MILKWEED

Purple-flowered milkweed thrives in the sandy or gravelly soils of the western Edwards Plateau and the Trans-Pecos areas of Texas. The careful observer of its delicate burgundy floral parts will note the miniature star in the center of the surface of the gynostegium, or female structure. The leaves are attractively arrayed in regular ranks upon the stems. Cattle will eat this species of milkweed in overgrazed areas.

A view of the seedpod is found on p. 181.

For additional information on milkweeds, see pp. 38, 186, and 233.

ASTERACEAE (Sunflower Family)

Conoclinium coelestinum

MISTFLOWER, BLUE BONESET, WILD AGERATUM

Mistflower's name is based on the Latin root *coeles*, meaning "heaven" or "sky," which indicates that the flowers are blue, and so they are, with a tinge of purple thrown in. It normally is found in the wetter regions of Texas and even grows well in poorly drained areas.

Several species in this genus are called boneset, apparently as a result of an epidemic of dengue fever in the American colonies in the early 1800s. The fever induced such severe pain that it was referred to as "breakbone fever," but treatment with mistflower quickly brought the

fever down, and as a result the plant became known as boneset—without ever touching a real bone. Its perceived ability to reduce fever made it a staple household herb for Native Americans as well as for early North American settlers. Native Americans also used it to get rid of parasitic worms and to treat colds and flu, urinary disorders, venereal diseases, snakebite, and epilepsy.

Mistflower is highly regarded as a landscape plant because it grows well in both shade and sun.

Liatris punctata var. *mucronata* (*L. mucronata*)
GAYFEATHER, BLAZING STAR

Gayfeather is a plant of the native prairies and therefore is generally not found in disturbed areas. It is noted for being drought tolerant, with a deep root system that carries it through dry years. Plants may survive more than five years. They grow well in limestone and almost all other soils except heavy clay.

Gayfeather's corms, or small, fiber-covered bulblets, are its means of propagation and also serve as a source of food and medicine. The corms contain inulin, a fructose polysaccharide that digests slowly and is useful as food for people with diabetes. The pounded corms, which are said to have a balsamlike or carrotlike flavor, were eaten by the Sioux as an appetite stimulant. The edibility of the corms varies by season—sweeter in the spring and tasteless at other times.

The Blackfoot Indians and Apaches used gayfeather corms to care for wounds and snakebite, and a tea from the root was used to treat bladder infections, diarrhea, and other abdominal ailments. The entire plant was powdered to make a tea for heart ailments. A blend of equal parts of crushed corms and a sweetener, such as honey, was used as a cough syrup. The corms can also be roasted and the smoke inhaled as a treatment for headache or nosebleed. In early times, New Englanders treated gonorrhea with gayfeather. There are some in the Native American and Mexican American cultures who believe that carrying or wearing a gayfeather corm with a cross carved in the inner surface will protect them from rattlesnakes or evil spirits.

Gayfeather, Blazing Star

Both domestic and wild animals feed on this plant, but it is not highly regarded for its forage value. Butterflies find it an important source of nectar, however.

Liatris has been domesticated for use in the home garden, where it adds color in the spring or fall. It is also produced commercially for both fresh and dried floral arrangements. The clusters of dried flowers on the *Liatris* spike evoked quite another image to the Sioux people, whose name for the plant referenced the spikes' similarity to clumps of deer manure.

A close-up of the flowers is found on p. 181.

Lygodesmia texana

SKELETON WEED

The genus name of skeleton weed picturesquely describes its slender, almost bare stems as a bundle (*desme*) of pliant twigs (*lygos*). Most plants rely on their foliage for photosynthesis, but because skeleton weed has only minuscule leaves, its green stems fulfill that role.

At first sight the flowers and the stems resemble those of chicory, but chicory stems have more visible and scruffy leaves along the stem, and the flowers are blue.

Native American children collected yellow balls of resin from the stems to chew, perhaps for the effect of the bright blue–colored saliva that develops as the gum is chewed.

A somewhat bluish tea may be made from the whole plant. This was given to nursing mothers to stimulate milk flow, based on the Doctrine of Signatures, a centuries-old traditional belief system, which holds that a trait of a plant gives an indication of its intended use in the human body; in this case, the milky sap was thought to encourage milk production. Nursing mothers were said to experience a mild euphoria after drinking the tea, and their children were believed to be healthier.

Skeleton weed may be poisonous to animals and humans because it can accumulate toxic levels of nitrates.

A close-up of the mature flowering head is found on p. 181.

Machaeranthera tanacetifolia

TAHOKA DAISY, TANSY ASTER

The heavyweight genus name for this hardy Texas native refers to the swordlike (*machaer*) appendages on the anthers. The species name indicates that the leaves are finely divided like those of tansy.

Tahoka daisies make a stunning purple carpet across pastures or fields. They were first commercialized from a field in Tahoka, Texas, south of Lubbock, several decades ago and have enjoyed increasing popularity among landscapers and flower enthusiasts everywhere.

Tahoka Daisy, Tansy Aster

The growth habit of Tahoka daisies is similar to that of yellow spiny daisies, with whom they share the genus name. They bloom continuously from spring to heavy frost and support a full cycle of flowers and seeds at the same time. Most species produce flowers and then go to seed in distinct stages, but *Machaeranthera* does everything at the same time, giving the plants a well-worn, frowsy look by midsummer.

The Hopis used *Machaeranthera* species as a stimulant for women in the throes of childbirth, and the Zuñis used them as an emetic to relieve upset stomach. Other Pueblo Indians chewed the fresh flowers to relieve stomachache. In keeping with the common Native American method of coping with colds and flu, the Navajos used the pulverized root to induce sneezing to relieve nasal congestion.

Vernonia marginata

PLAINS IRONWEED

Vernonia was named after the English botanist William Vernon, who made collections in North America in the late 1600s. The name "ironweed" refers to the toughness of the stem.

Vernonia's violet flowers are a pleasant contrast to the more common "yellowscape" of the Texas plains flora. Plains ironweed is drought tolerant as well as inviting to butterflies, whether in pastures, along roadsides, or in desert gardens.

The plants are often found in large colonies because they reproduce by underground stems, or rhizomes. These rhizomes were the source of several medicines important to the Cherokee people, who used them to reduce pain in childbirth and to prevent menstruation for two years after childbirth. This tribe also used the roots to treat stomachache and bleeding, as well as to make a mouthwash that improved gum health for those with loose teeth.

Where ironweed grows in pastures, its unpalatable chemical content protects it from grazing animals. *Vernonia* seeds contain significant quantities of vernolic acid, an important oil with a potential market in paints, pharmaceuticals, and other industrial applications.

COMMELINACEAE (Dayflower Family)

Tradescantia hirsutiflora
HAIRY SPIDERWORT, SNOTWEED

The genus name *Tradescantia* was given in honor of John Tradescant and his son John, Jr., who were the royal gardeners during the reign of Charles I of England in the 1600s. They planted material in the king's gardens that they collected during their own expeditions, as well as plants and seeds sent to them from America and elsewhere. The spiderwort was among a number of New World species that they introduced to England, and it is still a common plant in many British gardens today.

For such a wonderful flower, *Tradescantia* has some truly unattractive names, such as "hairy spiderwort" and "snotweed," a name conferred on it because of its mucilaginous sap.

This lovely plant puts on a show during the late spring and summer. The flowers have three petals, in shades of blue, purple, pink, or white. The buds and blooms are clustered at the top of erect stems.

Hairy Spiderwort, Snotweed

Native Americans were so taken by spiderwort's lovely appearance that young men of the Dakota people would sing love songs to *Tradescantia* flowers in the absence of their beloved because of the flowers' beauty.

The leaves and flowers are nutritious and can be eaten raw in a salad or as a cooked vegetable. The flavor is fine, but some people are put off by the slimy texture. The bitter-tasting root should be avoided, as it contains saponins, soapy chemicals that have useful medicinal properties.

FABACEAE or Leguminosae (Bean or Pea Family)

Astragalus mollissimus

PURPLE LOCOWEED, WOOLLY LOCO, RATTLEWEED

Among the little tarsal bones in the foot is the ankle bone, known as the astragalus bone, which connects the other tarsal bones to the fibula and tibia, the leg bones. The astragalus bones of deer and other mammals are just the right size and shape (four-sided) to use as dice for gaming

or for divination. Indeed, a number of cultures around the world "shake the bones" or "throw the bones" for just these purposes. When Linnaeus named the genus *Astragalus*, he may have based it on the similarity of *Astragalus*'s kidney-shaped seeds to the curvature of the ankle bone, or possibly because the sound of the seeds rattling in the pod was reminiscent of "shaking the bones."

The genus *Astragalus* is huge, with well over fifteen hundred species globally, and is found throughout the western United States. Purple locoweed is found on rangeland throughout the Central Plains from the Dakotas to Texas and is one of the most toxic in the genus.

Many *Astragalus* plants contain a toxic alkaloid called swainsonine that permanently damages the nervous system of grazing animals and causes them to act *loco*, the Spanish word for "crazy." Animals can even become addicted to locoweed, and the poison can be passed through the milk to nursing young. Affected animals will tremble and carry their heads low while staring vacantly. Locoweed is particularly dangerous to horses, for which only a small amount may be fatal. Horses that have recovered from the effects of the plant may be permanently affected and unsafe to ride.

Research in 2004 at New Mexico State University with *Oxytropis*, a cousin of *Astragalus*, indicates that the toxic alkaloid may actually be produced by a fungus growing in association with the plant rather than produced directly by the plant itself.

Purple locoweed and other members of the *Astragalus* genus accumulate minerals, such as selenium and barium, from the soil in which they grow. These heavy metals play a role in causing the many negative symptoms that result from consumption of the plant. Selenium poisoning, also called "blind staggers," will cause such symptoms as weight loss, disorientation, rough fur, chewing on fences, and even hemorrhaging, blindness, and paralysis. *Astragalus*'s ability to tolerate selenium and barium and to accumulate them in its tissues makes it of potential use in reclaiming land contaminated by excess selenium.

Native Americans used the root of some species of *Astragalus* as food, and in some cases, as an immune stimulant or as a spice. A number of butterfly species utilize *Astragalus* as a food source.

A view of the seedpods is found on p. 181.

For a comparison of *Astragalus* with *Oxytropis*, please see p. 160.

For other photos of *Astragalus*, see p. 255.

Dalea lasiathera (Parosela lasiathera)

PURPLE DALEA

The genus *Dalea* is named for the seventeenth-century English botanist, physician, and pharmacologist Samuel Dale. There are about two hundred species, primarily found in Mexico and the southwestern United States. *Dalea* is usually found on the limestone hills and open areas of the Edwards Plateau, the Rio Grande Plains, and the Trans-Pecos. Most species not only offer high-protein forage for wildlife but also make good perennial ground cover in xeriscapes.

Medicago sativa

ALFALFA, LUCERNE

Alfalfa is a native of North Africa and the Middle East, where the Arabs still feed it to their horses to increase their strength and speed. "Alfalfa" is the Spanish version of its Arabic name. It is believed to be the oldest forage crop introduced to North America and is cultivated as a valuable food crop for both livestock and humans. It now grows widely across Texas as an escape from cultivation. Alfalfa crops enrich the soil because of the nitrogen-fixing bacteria that live in nodules on their roots.

Europeans have long advocated consumption of a flavorful tea brewed from the leaves as a treatment for chronic illnesses. Natural healers recommend alfalfa tea as a soothing drink during recuperation from infections, particularly since it is thought to be beneficial if one is taking antibiotics. Alfalfa tea is a home remedy for the prevention of allergies. Because the leaves contain vitamin K, which aids in the clotting of blood, herbalists have prescribed alfalfa for hemorrhage, anemia, and menstrual difficulties. A leaf poultice has been used to treat earaches.

High in protein, minerals, and vitamins, alfalfa sprouts make a healthy addition to a salad, and the young leaves make a tasty cooked vegetable. However, people with autoimmune diseases, such as lupus and rheumatoid arthritis, should avoid consuming alfalfa, especially the sprouts and seeds.

Early American settlers believed that hanging green sprigs of alfalfa in their bedrooms would reduce the number of bedbugs, with which they were continually plagued. A yellow dye can be produced from the seeds, and the stem fibers have been used in making paper. The oil from the seeds can be used in the manufacture of paints.

Oxytropis lambertii

LAMBERT CRAZYWEED, LAMBERT LOCO, STEMLESS LOCO

Oxytropis means "pointed (*oxy*) keel (*tropis*)," referring to the sharp beak on the two lower petals. The handsome appearance of Lambert crazyweed in flower belies the danger and possible death that lurks within. It is regarded as one of the most dangerous and economically costly plants in the western United States because of its harmful effects on grazing livestock.

Like its cousins in the genus *Astragalus*, *Oxytropis* contains alkaloids that cause chronic neurological damage to grazing animals. If adequate forage is available, animals will avoid *Oxytropis* and *Astragalus*, but in the spring, herds should be withheld from pasturelands containing locoweed until the grasses are well established. Animals must consume the plants for two weeks to a month before symptoms of damage to the central nervous system appear. These symptoms include depression, agitation, poor muscle coordination that results in staggering, poor depth perception, and possibly death. Partial recovery is possible if animals are prevented from continued grazing. However, once animals acquire a taste for locoweed, they become addicted and will seek out the plants.

The toxic action is related to inhibition of the enzyme mannosidase, which results in accumulation of the sugar mannose in the lysosomes of cells, where the sugar would normally be metabolized. This lysosomal storage disease is reminiscent of Tay-Sachs disease in humans, which is a genetic disorder resulting in accumulation of certain lipids in lysosomes that also cause damage to the central nervous system.

Because eating the plants resulted in characteristic symptoms of locoism, it was long assumed that the plant itself produces the offending alkaloids. However, recent research indicates that a fungus living within *Oxytropis* may be the actual agent responsible for producing the toxic alkaloid. This would account for reports of variations in toxicity from one region or one season to another and would also explain why *Astragalus* and *Oxytropis* can be safely eaten in certain areas.

Views of the seedpods and whole plant are found on p. 181.

For a comparison of *Oxytropis* with *Astragalus*, please see pp. 156 and 160.

Psoralidium tenuiflora (Psoralea tenuiflora)

SCURFY PEA, SLIMFLOWER SCURFPEA, MANYFLOWER SCURFPEA

Scurfy pea resembles domesticated alfalfa, but the stems are weaker, making it more sprawling, and the leaves are narrower and less numerous. With roots penetrating ten feet into the soil to tap into moisture, this plant is very drought resistant. Livestock aren't particularly fond of its flavor, but they will eat it, especially when it first begins to grow. There have been some reports of scurfy pea poisoning in pigs, however.

At the end of its growing season, scurfy pea's dry stem snaps off in the wind, and the entire plant rolls across the prairie, dispersing its seeds in the same manner as tumbleweeds (Russian thistle, *Salsola tragus*).

The Lakotas brewed a tea from the roots to allay the pain of headaches, and they also burned the plant to keep mosquitoes at bay. The Dakota people wove the plants into a sun visor to shade their heads during the hot days of summer. Even today, herbal practitioners make many health claims for scurfy pea, saying that it is effective in treating eczema and hair loss as well as in fighting infections. They also recommend it for promoting bone health.

Sophora secundiflora
TEXAS MOUNTAIN LAUREL, CORAL BEAN, BIG DRUNK BEAN

Texas mountain laurel is a slow-growing shrub or small tree, and one that is well worth the effort to cultivate in the garden. Its clusters of rich purple flowers are reminiscent of wisteria and have a delightful smell very much like the fragrance of grape Kool-Aid. It does well in sun or light shade in any well-drained soil, needs little irrigation once established, and is resistant to pests and diseases. If conditions are right, Texas mountain laurel can attain heights slightly in excess of thirty feet.

A gardener who wants to start with seeds will have to either rob the cradle and take the seeds from the bean before they have formed their tough red seed coat or scarify the older beans. The most effective scarification method is to soak the seeds in sulfuric acid for about fifteen minutes prior to planting, but simply nicking the seed coats with a file will work.

Texas mountain laurel contains several alkaloids that may be toxic to ruminants and other mammals. The poisonous compounds are more potent in mature leaves and fruit as well as the seeds, but if the seeds are not crushed, they will pass through the digestive tract without causing

any harm. Affected animals show muscle tremors, move stiffly, or fall, which is doubtless where the name "big drunk bean" originated, but most animals will eventually recover.

The red seeds have been used by Native Americans as a ritual hallucinogen, although the powerful alkaloids are potentially fatal. At least a dozen tribes practiced the Red Bean Dance in which visions were sought (and generally found) after the participants consumed a drink prepared from the seeds. Ceremonial necklaces made from the brick-red seeds were worn in honor of the occasion. In addition to bringing about hallucinations, the toxins may cause nausea, convulsions, and paralysis of the diaphragm, which leads to asphyxiation.

Seedpods and seeds are shown on p. 182.

GERANIACEAE (Geranium Family)

Erodium texanum

TEXAS STORK'S BILL, DESERT HERON'S BILL

Stork's bill is named for its unfriendly-looking fruit, which is a pointed structure resembling the beak of a bird, such as a stork or crane. Appropriately, the family name, Geraniaceae, means "crane." Each seed's "beak" is actually the female structure (style) through which the pollen tube passes to deliver the sperm to the egg. This long, pointed structure responds to the changes in humidity by coiling and uncoiling and corkscrewing itself into the ground—an ingenious and cunning device for planting the seed in the soil when there is adequate moisture for germination.

The plants overwinter as a rosette, or cluster of leaves growing close to the ground, and send up flowers at the slightest hint of spring. In balmy locations stork's bill may bloom even in January. The leaves are

edible as a cooked green vegetable or chopped up in a salad. The cooked root is said to taste something like a turnip.

Stork's bill is a traditional health remedy in the southwestern United States. For its various medicinal uses, however, a large amount is required to bring about the desired effect. A tea from the leaves provides a gentle diuretic to alleviate water retention during the menstrual cycle as well as to relieve painful and heavy menstruation. Native American women relied upon it to reduce bleeding after childbirth. Stork's bill is also useful for bladder and urethral infections. The astringent properties of stork's bill tea are helpful for constricting soft tissue, which makes it efficacious as a wash for acne and skin abrasions or for relieving diarrhea. Folk medicine indicates its use in a bath for arthritis relief, and Native Americans used the root to blunt the pain of toothaches.

Domestic and wild grazing animals find stork's bill palatable and benefit from the high protein content.

A view of the seedpods of Texas stork's bill is found on p. 182, along with the flowers and seedpod of a related species.

HYDROPHYLLACEAE (Waterleaf Family)

Phacelia integrifolia
GYPSUM PHACELIA, GYP BLUE CURLS, SCORPION-WEED

It may seem surprising that *blue* curls are in the purple section. In fact, there are some "blue" flowers that look purple, and some "purple" flowers that look blue . . . or even pink! It just proves that color, like beauty, is in the eye of the beholder.

Phacelia integrifolia, *gypsum phacelia*

Phacelia congesta, *blue curls*

Blue curls have coiled inflorescences that unfurl as the bluish-purple flowers make their appearance a few at a time. From each bell-shaped blossom, anthers extend far beyond the petals, giving an airy, lacy look to the floral group.

The genus name *Phacelia* comes from the Greek word for "cluster," in reference to the flowers. This particular type of inflorescence is called a scorpioid cyme, because of its similarity to the coiled tail of a scorpion. One of the species is called a scorpion-weed for this reason.

All types of blue curls are easily cultivated; they make a pleasing visual display in the garden and attract butterflies as a bonus. Cattle and white-tailed deer eat the plants.

Some species of *Phacelia* have been used for food and medicine. Leaves of blue curls can be boiled and eaten as a cooked vegetable, and the roots and leaves have been used to treat rashes, sprains, and swellings.

The robust stems and leaves of gypsum phacelia are shown above, along with a close-up of the flowers of a related species, *P. congesta*.

LAMIACEAE (Mint Family)

Lamium amplexicaule

HENBIT, DEADNETTLE, GIRAFFE HEAD

Henbit can bloom year-round, but it is most noticeable in the cold months when there is no competition. It prefers disturbed soil, and it is often found around homes. Many (undiscerning) people think of it as a weed.

Henbit has been used to treat the soreness of rheumatism, to lower a fever, and to act as a laxative when the need arises. The young leaves may be added to a salad or cooked as a green vegetable.

Monarda citriodora

HORSEMINT, LEMON MINT, LEMON BEEBALM

The name *Monarda* was given in honor of Nicholas Monardes, a Spanish physician in the 1500s who described the medicinal attributes of several New World plants; *citriodora* refers to the lemon or citrus fragrance of the plant.

This tall, handsome plant often forms large colonies across the countryside, especially in the Texas Hill Country. Horsemint's good looks and long bloom period (spring through fall) make it a popular choice for landscaping. As a bonus, bees, butterflies, and hummingbirds frequent gardens that feature *Monarda*.

The leaves may be added raw to a salad, or they can be cooked as one would any other vegetable. A refreshing herbal tea can be brewed from the leaves. The Hopis cooked lemon mint with rabbit, and other Pueblo Indians used it to flavor beans and stews. They also dried it for winter use. The Tewas ate it as cooked greens. In addition to its culinary properties, lemon mint and other *Monarda* species make soothing teas for sore throats and cough. Native American women used this plant to relieve menstrual pain.

The citrus fragrance of lemon mint is a result of an oil called citronellol, which is used commercially in perfumes and insect repellents.

Scutellaria drummondii

SKULLCAP

One would expect this pretty little plant to have a more appealing name than "skullcap," but when a person examines it in detail, it becomes clear why the name is appropriate. Right above each bloom is a cup-shaped appendage of the sepals, which resembles the skullcaps worn by monks and by Jewish men.

This low-growing perennial makes a great ground cover in a native plant landscape. Skullcap requires very little maintenance, and it is drought tolerant.

Unlike many members of the mint family, the bitter-flavored skullcap is not edible. As a folk remedy, it has been used to treat nervous disorders, convulsions, and headaches. Some traditional healers have used this plant to help barbiturate addicts and alcoholics overcome the difficulties of withdrawal. There is some indication of its value as an antiviral agent as well as a chemical enhancer to boost the effectiveness of chemotherapy for lung cancer.

NYCTAGINACEAE (Four O'clock Family)

Mirabilis glabrifolia

FLAT-TOP UMBRELLAWORT

Mirabilis means "wonderful," an appellation that would probably make a person viewing this plant shrug in confusion. It looks okay, but not exactly "wonderful." For one thing, a person would have to get up really early to get a good look at the flat-top umbrellawort while it is still bearing its pink or purple blooms. The floral parts quickly fall away, leaving behind a quaint seed-bearing structure resembling a little umbrella. As the photos on p. 182 show, tiny fruits called anthocarps nestle inside it.

In the Far East and in the tropics, members of this genus have supplied various types of medicine, provided some of the constituents in cosmetics, and have been a source of the dyes used to color jellies. Maybe it is "wonderful" after all.

A close view of the mature fruit and leaf arrangement is shown on pp. 182–183.

PASSIFLORACEAE (Passionflower Family)

Passiflora incarnata

PASSIONFLOWER, MAYPOP, GRENADILLA

When the Spanish priests came to the New World, they used any means available to explain the story of Jesus to the Native Americans. Imagine how pleased the priests were to discover the gorgeous maypop, in which they saw an amazing number of Christian symbols. Using the bloom as a visual aid, they explained such concepts as the Trinity, using the way the one stigma is made of three parts or the manner in which each leaf is divided into three lobes. The five stamens represented the five wounds in the crucified Savior, while the sepals and petals were symbolic of the ten

Passionflower, Maypop, Grenadilla

faithful apostles, not counting Judas the Betrayer and Peter, who denied the Lord. Jesus' crown of thorns is symbolized by the ring of long filaments inside the petals, and the royal purple color shows he is the king. The plant is a vine with clinging tendrils, demonstrating the way God will support his children if they hold on to him. The priests made such use of this lovely flower that it came to be known as the passionflower, since it told of the Passion (the death, burial, and resurrection) of Christ.

Native Americans were already quite fond of the passionflower as a source of both food and medicine. This particular species has an orange-yellow fruit that is rich in vitamin A and niacin and may be up to two inches long, the largest of the fruits of any passionflower. The tasty juice around the seeds can be sucked from the fruit, squeezed and used as cold drink, or made into jelly.

Passionflowers have long been used to treat insomnia and to soothe pain. In recent years passionflower has won approval in Germany as an herbal medicine for treating nervous conditions. The juices of the plant are antibacterial and are used to tend burns, wounds, inflamed eyes, and irritated skin. The fruits are also thought to lower blood pressure.

Naturally, so beautiful a vine is a popular choice as an ornamental. A word of caution: it may thrive to the point of taking over the garden, as it scrambles over small plants, bushes, trees, and fences. In fact, the weight of its vines can cause fences to collapse.

SCROPHULARIACEAE (Snapdragon Family)

Leucophyllum frutescens

CENIZO, LIAR'S BUSH, TEXAS PURPLE SAGE

Experienced old-timers believe that when cenizo blooms, rain is just a couple of days away. Indeed, this plant does respond to atmospheric changes that accompany actual or potential rainfall by bursting into bloom with a profusion of pink to purple flowers. It thus offers a note of hopeful cheer to rain-starved West Texans, but sometimes it cruelly disappoints when rain doesn't materialize, earning it the epithet "liar's bush."

The genus name *Leucophyllum* literally means "silvery-white leaves," which is an apt description of cenizo's appearance. Gardeners have fallen in love with cenizo's delightful combination of silver foliage and purple blooms, and it is now commonly used in landscapes in the western half of the state. Cenizo's preferred soil is limestone or caliche, which is also found in abundance in that region of Texas. Browsing wildlife and livestock find it a nutritious forage plant, so gardeners should evaluate the potential problems with animals in their area before including this plant in their plans. Cenizo tea, made from the leaves and flowers, has a pleasant taste and may be enjoyed for its own sake or for its medicinal value. The mildly sedative tea is a folk remedy for colds or flu, stimulating sweating that breaks a fever.

A close-up of the flowers is found on p. 183.

Penstemon cobaea

COBAEA BEARDTONGUE, WILD FOXGLOVE, LARGE-FLOWERED BEARDTONGUE

Cobaea beardtongue and other penstemon species are not related to the true foxglove (*Digitalis purpurea*), even though some of the common names may lead one to think so.

The flowers of this showy species, which is found throughout Central Texas to the Gulf, vary from mostly white, to deep pink, to lavender. With blooms covering more than half the stem, it is an eye-catching feature in a native landscape. It is also attractive to moths, hummingbirds, and butterflies and serves as the larval host to the dotted checkerspot butterfly.

For more information about *Penstemon* characteristics and medicinal properties, see p. 220.

Penstemon fendleri

PURPLE FOXGLOVE, FENDLER'S PENSTEMON

The foxglove got its common name from its role in an old children's story about a fox who wanted to sneak up on some chickens. His toenails kept clicking and alarming the birds before he could make a meal of them, so he deviously slipped the blooms of this flower over his paws like gloves and was able to raid the henhouse quietly.

Purple foxglove is used in native landscapes, where its showy, delicately lavender inflorescences will brighten the landscape from April through June. Thriving in well-drained, calcareous soil with access to plenty of sunlight, plants can exceed two feet in height.

For a close-up of the leaves of purple foxglove, see p. 183.

For more information about *Penstemon* characteristics and medicinal properties, see p. 220.

SOLANACEAE (Nightshade Family)

Quincula lobata
PURPLE GROUND CHERRY

A colorful, low-growing bouquet, this plant is a pleasant surprise in a dry, parched area where it seems that everything else has withered and blown away with the Texas winds. Because it is so drought resistant, it makes a superb addition to a rock garden.

The Kiowa people who lived in what is now Kansas used the pounded root to dress wounds and to make a tea for severe coughing. For a swollen sore throat, fully ripe fruit can be used as a compress. Its sweet yellow berry is used to make preserves as well as salsa verde.

A view of the seedpods is found on p. 183.

Solanum eleagnifolium
SILVERLEAF NIGHTSHADE, WHITE WEED

Silverleaf nightshade is a perniciously troublesome weed because of its extraordinary skills in propagating itself by underground rhizomes, especially in cultivated fields or in gardens. Mechanical destruction, such as hoeing, is almost sure to fail, and herbicide control succeeds only with persistent and diligent efforts.

It also is highly toxic, thanks to the glycoalkaloid solanine found in all parts of the plant at all stages of growth, but especially in the ripe fruit. Solanine damages the nervous and digestive systems of cattle, but sheep and goats are not seriously affected by it.

Although the berries should never be eaten on their own, Native Americans and Mexicans make a delicious white cheese, called asadero, using the crushed yellow fruits of this plant to coagulate the milk.

Those skilled in the use of herbal remedies have used the berries to treat toothaches and sore throats. A salve was made from the crushed fruits to bring soothing relief from poison ivy rashes. Zuñi Pueblo people suffering from tooth cavities chewed the root and put it in the hole to relieve the pain. The Kiowas tanned the hides of animals with a mixture of the animals' brains and silverleaf nightshade berries.

A view of the seedpods is found on p. 183.

UMBELLIFERAE or Apiaceae (Parsley Family)

Eryngium leavenworthii

ERYNGO, FALSE PURPLE THISTLE, PURPLE PINEAPPLE

Few people would guess that eryngo is not a thistle but a member of the carrot (or parsley) family. Eryngo grows in prairies from Texas to Kansas and is also found in disturbed soil or "abandoned" areas. It thrives in alkaline, calcareous soil.

There is some speculation about the derivation of the genus *Eryngium*. Because this plant is believed to relieve gas in the stomach, the name is said to come from the Greek word *eruggarein*, which means "to eructate"—a polite way of saying "to burp." On the other hand, it may be derived from *erungos*, meaning "thistle." The species name was given in honor of Dr. M. C. Leavenworth, a botanist and surgeon who lived in the early 1800s.

This much-branched species, which may grow to be as tall as a person, makes the late-summer countryside look purple from a distance. What appears to be a single flower is actually a closely grouped cluster of tiny blooms with spiny bracts festooning both the top and the bottom of the head. As if the purple weren't enough, eryngo completes the special effects with a lacy sprinkling of blue stamens. Even the stems forgo the typical green and opt for silver or plum-colored pigments.

Because eryngo maintains its shape and color well, it makes a unique contribution to a dried flower arrangement, adding textural contrast as well as visual interest.

Eryngo species have been used for at least two thousand years for the relief of both pulmonary and urological disorders. Extracts of some species of eryngo have been used to alleviate the pain of scorpion stings. In Europe and the Middle East, the roots were candied and marketed as an aphrodisiac.

VERBENACEAE (Verbena Family)

Glandularia bipinnatifida (*Verbena bipinnatifida*)

PRAIRIE VERBENA

Verbena is derived from a Celtic word meaning "to drive the stone away," referring to elimination of kidney stones, for which this plant was once used.

The three verbena species pictured here have very similar flower structures, reminiscent of a gingerbread man that has fallen into the purple dye vat, but the plant appearances are very different. Prairie verbena, shown in the main photo, spreads horizontally, with the flowers rising in clusters above the "mat." Prostrate vervain, shown on the next page, also has a horizontal growth habit, but the flowers are far less showy. Slender vervain, appearing alongside it, stands in perpendicular contrast to its

Verbena bracteata, *prostrate vervain*

Verbena halei, *slender vervain*

cousins, with delicate stems that dance in the breeze above a few basal leaves.

The *Verbena* genus has been highly regarded as a curative plant group for almost every ailment known to humankind. Traditional healers recommend a weak tea made from the leaves and flowers to use as a sedative, as a treatment for colds, and as a therapy for deep bruises. If the tea is too strong, it may cause nausea, and in fact, when an emetic is called for, the tea is purposely allowed to steep a longer time than usual. Verbena is thought to be especially good for colds and flu because it breaks fever by inducing sweating, and it encourages expectoration of the phlegm left over from the immune system's war with the viruses. Midwives have used vervain in dealing with difficult childbirths because they feel that it gently encourages uterine contractions and reduces bleeding. Native Americans used vervain to increase milk flow in nursing mothers, and modern chemical analysis has shown that the plant contains a glycoside, verbenin, that stimulates breast milk production.

Beliefs dating from the time of the Roman Empire and even those of European Druids held that all vervains possessed magical powers. Anything from preventing headaches to keeping away witches and evil spirits was attributed to these plants. They were carried for good luck in general and for attracting wealth in particular. Soldiers used to carry verbenas with them to prevent injury. Many species of verbena have been used in love charms of all sorts, and German brides have a romantic tradition that wearing vervains in their hair on their wedding day will ensure a lasting, happy marriage.

Vitex agnus-castus

VITEX, CHASTE TREE, MONK'S PEPPER

Vitex is a tall, drought-tolerant shrub and is much favored as an ornamental in the southwestern United States. The name "chaste tree" reflects its use since ancient times as a means of suppressing sexual desire. The young women of ancient Athens would make chains of the leaves as a sign of purity during certain sacred celebrations. Catholic monks chewed vitex leaves to help them keep their vows of chastity. Although the genuine effectiveness for this purpose may be questioned, vitex is known to contain several reproductive hormone derivatives in the bark, leaves, and flowers, perhaps giving some credence to the belief that it diminishes ardor.

Slightly more convincing are claims for its value in treating a wide range of female disorders. After a study in Germany showed its efficacy in treating premenstrual syndrome, vitex is being marketed as a treatment for this debilitating malady. Vitex has also been used as an aid in passing the afterbirth and in controlling bleeding following childbirth. Its use should be avoided during pregnancy. The berries have been used to relieve pains and weakness in the limbs.

VIOLACEAE (Violet Family)

Viola sororia (*V. missouriensis*)
MISSOURI VIOLET

Walking the trails alongside some of Texas' rivers in the early springtime, one may come across a delightful little flower, the Missouri violet, which prefers moist, shady areas. The soft lilac petals have dark purple lines that may look like decorations to humans but actually are markings that insects follow to locate nectar.

Native Americans learned to put crushed violet leaves on boils and swellings to ease the pain and promote faster healing. The plant's juices loosen congestion and were given in a syrup for coughs. The Ojibwas made use of violet's roots for bladder problems and for sore throats. They used the whole plant to make a tea to strengthen the heart, and they lowered fevers with leaf extracts. Violets have sedative and laxative properties as well.

All the aboveground parts are edible, and the leaves contain particularly high levels of vitamins A and C. One should never eat the root of a violet, because its emetic properties might upset the stomach.

Leaf arrangement and fruiting structures are important not only in distinguishing one plant family from another but also for identifying individual species. The fruits are almost as enjoyable as the flowers, with their ingenious strategies for dispersing seed.

Asclepias brachystephana, *purple-flowered milkweed; note the long linear leaves and opened seedpod with seeds and fluff emerging—with spider in residence, p. 150.*

Liatris punctata *var.* mucronata, *gayfeather; close-up of flowers with imbricated phyllaries enclosing each cluster, p. 151.*

Lygodesmia texana, *skeleton weed; mature flower head with achenes, each containing a single seed, and with fluffy pappus for wind dispersal, p. 152.*

Astragalus mollissimus, *purple locoweed; seedpods with filamentous style still attached and with dense hairs on plant, p. 156.*

Oxytropis lambertii, *Lambert crazyweed; mature, dehisced (opened) seedpods, p. 160.*

Oxytropis lambertii, *Lambert crazyweed; field appearance, p. 160.*

Sophora secundiflora, *Texas mountain laurel; bean pods, p. 162.*

Sophora secundiflora, *Texas mountain laurel; red seeds, p. 162.*

Erodium cicutarium, *stork's bill; close-up of flowers, p. 163.*

Erodium cicutarium, *stork's bill; note the deeply incised leaves and resemblance of seedpods to the head of a stork, p. 163.*

Erodium texanum, *Texas stork's bill; field appearance with seedpods and leaves that are somewhat heart shaped and palmately veined, p. 163.*

Mirabilis glabrifolia, *flat-top umbrellawort; mature anthocarps, each containing a seed, p. 168.*

Leucophyllum frutescens, *Texas purple sage; close-up of flowers, p. 171.*

Mirabilis glabrifolia, *flat-top umbrellawort; opposite, clasping leaves, p. 168.*

Penstemon fendleri, *purple foxglove; note blue-gray waxy appearance of stem and leaves, which are opposite with bases that clasp the stem, p. 173.*

Quincula lobata, *purple ground cherry; seedpods with inflated calyx, p. 174.*

Solanum eleagnifolium, *silverleaf nightshade; mature fruits, p. 174.*

Pink Flowers

ASCLEPIADACEAE (Milkweed Family)

Asclepias speciosa

SHOWY MILKWEED, LECHONES, COMMON MILKWEED

Growing up to four feet tall and displaying clusters of light pink flowers, this plant lives up to its common name "showy milkweed," which is echoed in its scientific name, *speciosa*, Latin for "showy."

As a rule, milkweeds should be considered poisonous, but broadleaved species such as showy milkweed are believed to contain less toxic amounts of glycosides and are thus not as nauseating as narrow-leaved species. Many tribes of Native Americans used showy milkweed as food, but they cooked it in several changes of water to remove the bitter compounds. They ate the young flowers, the large roots, and the immature seeds, and they boiled the flowers down to make sugar. The young seedpods, which contain a meat-tenderizing chemical known as asclepain, were cooked with buffalo meat. Leaving no part unused, the Native Americans fashioned spoons from the dried seedpods.

A tea derived from the root is valuable as a decongestant. Overconsumption, however, may result in nausea or vomiting. The Cheyenne people brewed an eyewash from the new leaves to treat snow blindness and other vision problems. The diuretic properties of the liquid extracted from boiled roots made it useful for treating mild kidney complaints and for relieving breast engorgement in nursing mothers. Several Native American tribes used the roots to induce temporary sterility.

A view of the seedpods is found on p. 224.

For additional information on milkweeds, see pp. 38, 150, and 233.

ASTERACEAE (Sunflower Family)

Acourtia nana

DESERT HOLLY

This small but enchanting plant was named in honor of Mary Elizabeth A'Court, a British amateur botanist of the 1800s. Desert holly's species name, *nana*, means "small" or "dwarf," indicative of its one- to six-inch height. The prickly, leathery, gray-green leaves resemble holly leaves. Finding it in flower is a special treat, since it so quickly transforms into a white puffball as the seeds mature.

As its name suggests, desert holly typically resides in the arid regions of the state, particularly in the Trans-Pecos region.

Carduus nutans

MUSK THISTLE, NODDING THISTLE, PLUMELESS THISTLE

Carolus Linnaeus gave the genus name *Carduus*, which is Latin for "thistle," to a number of thistle species in the mid-1700s. The name *nutans* refers to this species' large, nodding flower heads, which look like pink shaving brushes and have a faint odor of musk. The flower head may be as broad as the palm of a hand, and the entire plant may exceed six feet in height. The broad, purple-tinged bracts under the flower head are attention grabbing on their own.

Thistles have an eons-long reputation in Europe and the Mediterranean region as a symbol for poor agricultural management, because they prefer to grow on good soil that has been neglected. *Carduus nutans* immigrated to North America with the early European settlers and has succeeded in establishing an equally unsavory reputation in its new

Musk Thistle, Nodding Thistle, Plumeless Thistle

home. Thistles seed prolifically, and some also spread by rhizomes. This characteristic, together with their spine-edged leaves, degrade the value of the land that they infest.

A close-up of the spiky unopened bud is found on p. 224.

Centaurea americana

BASKET FLOWER, STAR THISTLE

Basket flower's genus name, *Centaurea*, originated with the myth that this plant was used to heal the wound that Hercules inflicted on a centaur's foot. The white to brownish bracts that cover the buds and subtend the flower head resemble basketwork (or an orderly phalanx of ticks or spiders). These bracts are called "phyllaries," and each tribe in the sunflower family has a typical shape and arrangement of phyllaries that helps in identification.

The basket flower's stamens are so sensitive to touch that they shoot out their pollen onto any insect that brushes against them. The purplish extensions around the edges of the flower head of the basket flower are not ray flowers, as they appear, but are tubular disk flowers.

Basket flowers perform well in both fresh and dried flower arrangements.

A close-up of the bracts under the flowering head is found on p. 224.

Cirsium horridulum
PURPLE THISTLE, HORRIBLE THISTLE, YELLOW THISTLE

The aptly named horrible thistle typically grows in wetlands and pastures from Texas to the eastern United States. Its presence indicates overgrazing and poor pasture management. The species name is Latin for "prickly" or "rough," and it is guaranteed to be given a wide berth by man and beast. The flowers are reported to be yellow in coastal areas and pink when farther inland. In spite of its rough nature, or perhaps because it is just so horrible, it has a number of devoted gardening fans and is available commercially in some nurseries.

Texas thistle, shown on the next page, has a less prickly personality than most of its thistle relatives, and both its young (preflowering) stems (peeled, of course) and roots make good cooked vegetables.

Yellowspine thistle, on the other hand, is classified as a noxious weed in California, because it spreads aggressively by seeds and rhizomes and reduces the value of rangelands. This species is also shown on the next page.

Overlooking the armaments and focusing instead on the flowers, it's easy to see their beauty—a perspective that butterflies share. They have no qualms at all about thistles' prickles and go straight for the nectar, which they find very attractive.

Cirsium ochrocentrum, *yellowspine thistle;*
note the yellow-tipped spines

Cirsium texanum, *Texas thistle; note the*
darker-tipped spines

Thistle tea was drunk by the Hopis to relieve colds and constipation; the Pueblo Indians from San Juan used the seeds to treat boils. The Zuñis used a tea made from the roots as a contraceptive. For the Navajos, the tea served as an eyewash, a sedative for headaches, and a febrifuge to control fever.

The stems of all thistle species can be processed for cordage—raw material for rope or other coarse binding products. Yellow dyes can be extracted from the entire plant.

Several hundred years ago, thistles played a role in saving Scotland from an invasion by the Vikings. As the Norsemen stepped ashore in the dark, they found themselves treading on thistles that the Scots had so thoughtfully laid out for their landing convenience. Alas, the Vikings were not tough enough to withstand the pain, and their cries alerted the Scots, who rushed out to trounce them. Needless to say, the Scots have a very high regard for thistles and subsequently immortalized them in song, made them the national flower, and formed a chivalric order of knighthood, the Order of the Thistle.

A close-up of horrible thistle is found on p. 224.

Echinacea angustifolia

PURPLE CONEFLOWER, KANSAS SNAKEROOT, ECHINACEA

Coneflower's genus name, *Echinacea,* is derived from the Greek word *echinos,* which refers to the flower head's spiny similarity to a hedgehog. The dried flower heads are quite stiff and have been used by Native Americans as combs.

There are at least six species of coneflower in Texas, and they frequently hybridize with one another.

Echinacea's pharmaceutical properties have given it an important position in the thriving dietary supplement industry in the United States and Europe because of its purported anti-inflammatory, antiviral, immune-stimulating, and other health-related characteristics.

American Indians made tea from the roots to use as a wash for wounds, as a gargle for sore throats, and even as a treatment for snakebite. The fresh root has been used to treat diphtheria.

Purple coneflowers are now available in many plant nurseries. They can be propagated by using rhizomes, or underground stems, or by separating off the "babies" that grow from the crown area, which is the part of the plant in contact with the soil.

A view of the whole plant and mature flowering head of *E. angustifolia* is found on p. 224.

Palafoxia sphacelata

SAND PALAFOXIA, RAYED PALAFOXIA

Palafoxia is named for the Spanish general José de Palafox, who fought against Napoleon in the early 1800s. The species name *sphacelata* means "appearing dead" or "having gangrene," clearly an incongruous appellation for this lovely, delicate flower.

Sand palafoxia, true to its name, thrives in the sand dunes of West Texas. It will readily adapt to the comforts of a home garden and makes an inviting profusion of lacy color in a landscape setting while also luring

Sand Palafoxia, Rayed Palafoxia

in the butterfly crowd. Sand palafoxia grows easily from seed, which should be sown in the fall for an early spring show.

A view of the flowering head is found on p. 225.

BIGNONIACEAE (Catalpa Family)

Chilopsis linearis

DESERT WILLOW, MIMBRE

The desert willow, a small, deciduous tree valued for its large, showy flowers and graceful form, is commonly used as a low-maintenance ornamental. Although its appearance is "willowy," it is not related to the true willows in the Salicaceae family.

Desert willow is a phreatophyte, or "water lover," which means that its roots reach down to tap into the water table. Its presence in the natural landscape indicates that the water table is not far below the surface.

Desert willow's flowers are fragrant and vary in color from white to pink or purple. It blooms throughout the summer, dropping its leaves in the fall but retaining its long, narrow seedpods through the winter.

A tea and hot poultice made from desert willow's flowers are purported to relieve coughing and pulmonary distress from colds and flu. The liquid extract also has strong antifungal properties and is used to treat yeast infections resulting from prolonged antibiotic or corticosteroid use. Stems and leaves can be pulverized and applied to ringworm or other fungal infections. This powder can also be dusted on scratches and stings.

A view of the whole plant is found on p. 225.

CACTACEAE (Cactus Family)

Echinocactus texensis

HORSE CRIPPLER, DEVIL'S HEAD

Sporting large, stout spines, the horse crippler cactus certainly earns its name. Not only can the spines injure a horse but they can also pierce a tire or the sole of a shoe.

Beautiful pink blooms with violet centers make this cactus an attention-getting addition to a rock garden. Furthermore, the local wildlife will appreciate horse crippler's presence, since the large flowers offer a feast of pollen and nectar for numerous insects, including bees. The red fruits will add a splash of color after the flowers have gone and provide tasty morsels for many types of birds as well as for Texas tortoises.

The sweet fruits are edible for humans as well. In Mexico the fruits are used in making candy as well as a chocolate-colored cheese. Native Americans have used the long spines as fishhooks.

A view of the whole plant with fruit is found on p. 225.

Opuntia imbricata

CHOLLA

The wine-colored blooms of cholla (pronounced "choya"), displayed over spiny branches, and its height—it can be taller than a person—make it a provocative landscape feature. The bright yellow fruits that follow the flowers persist through the summer and into the next growing period, providing year-round color. Its dried woody stems make specialty walking canes or a novel framework for flower arrangements.

The fruits are sweet and edible, and the young stems make a tasty fried vegetable. Southwestern Indians called the month of March the "cactus moon" because, with other food being scarce, they ate young cholla buds. These should be boiled until tender, and they are also good sautéed with yellow squash. Hopis used the roots to treat diarrhea and also made a tea for use as a hair tonic. They used the cholla's juice as a fabric stiffener.

A close view of the yellow fruit is shown here. The oddly branching growth habit is shown in a field view in the Exploring Further section, p. 225.

Opuntia imbricata, *cholla; fruit.*

CONVOLVULACEAE (Morning-Glory Family)

Convolvulus arvensis

FIELD BINDWEED, CREEPING JENNY, WILD MORNING-GLORY

Field bindweed is a native of Europe, and in its new American home, despite its lovely bloom, it is enthusiastically despised. Living up to its genus name, *Convolvulus*, which means "entwining," field bindweed wraps around the stems of other plants and robs them of moisture, nutrients, and light. It is difficult to eradicate because its rhizomes descend into the soil as far as fifteen or twenty feet, and it will grow back from any fragment left behind. In addition, its numerous seeds are viable for up to fifty years in the soil. Grazing animals can keep this weed under control to some extent, but there are reports that bindweed may be toxic, especially to pigs.

On the positive side, some research is under way regarding the plant's content of proteoglycans (PGM), a chemical that shows some effectiveness in shrinking tumors. It would take something as momentous as a cure for cancer to make some people lose their antipathy for this wicked weed.

Ipomoea leptophylla

BUSH MORNING-GLORY, MAN ROOT, BIG-ROOT MORNING-GLORY

As its name indicates, this morning-glory may become a bush of over three feet in height. Bush morning-glory thrives in all soil types, inhabiting the prairie from the Panhandle of Texas north through the Great Plains into South Dakota.

The root is the most intriguing part of the plant—it is enormous and can reach the size of a human. The Pawnee people burned this tuber for the effect of the smoke on nervous disorders and as an antidote for bad dreams. They also powdered it for dusting over the body to relieve pain. The root could also be consumed raw to alleviate stomach problems. Its most important medical value seems to have been in the treatment of urinary disorders. Great Plains Indians used the huge, fibrous root as a smoldering "storage unit" for fire. Apparently the root was not a culinary delight, however, since the Indians ate it only when little else was available.

The large, showy flowers and huge storage root make it a potentially valuable landscaping plant not only for home gardens but also for parks and roadsides.

For additional information on other morning-glory species, see pp. 139 and 249.

FABACEAE or Leguminosae (Bean or Pea Family)

Dalea formosa

FEATHER DALEA, YERBA DE ALONSO GARCIA

An encounter with feather dalea when it is not flowering is somewhat less than auspicious, but when it bursts into bloom in the early spring, feather dalea is a feast for the eyes. It also provides fresh green browse for deer and livestock.

Massed plantings of this small (two to six feet tall) shrub require little water or care and will reward the gardener with lovely purplish-pink blooms from April to October. A relative, the silver dalea, has very similar flowers, but without hairs.

The flowering branches may be brewed to make a delicious, slightly tart tea, which also has medicinal properties. Pueblo Indians and Apaches still use the tea to alleviate the pain in aching bones. Some soak in a fragrant, hot bath with feather dalea branches immersed in the water. For the most effective cure, they have a cup of the tea while they relax in the tub. The Hopi people recommend the tea as a treatment for the flu.

Feather dalea bark has been used as a source of dye for baskets.

Lespedeza virginica

SLENDER BUSH-CLOVER, SLENDER LESPEDEZA

Slender bush-clover grows up to three feet tall, bearing clusters of pink or purple (occasionally white) flowers interspersed with the trifoliolate leaves (compound leaves with three leaflets). Blooming from summer into fall, most of the flowers are pollinated by bees, but the flowers are

Slender Bush-Clover, Slender Lespedeza

capable of self-pollination. Other insect visitors include various butterfly species whose caterpillars feed upon the foliage.

Preferring sandy, rocky, or loamy soils, this perennial requires sunny, open areas to establish seedlings, which do not compete well with other plants. Frequent disturbances, or even fires, encourage the establishment of colonies of this soil-stabilizing, deep-rooted legume.

Slender bush-clover is a favorite fare for deer, and it also provides food for livestock, rabbits, groundhogs, and other herbivores. The seeds are eaten by many birds, notably the bobwhite quail and wild turkey. The Comanches are said to have made a beverage from leaves of other *Lespedeza* species.

Mimosa nuttallii (Schrankia uncinata)

SENSITIVE BRIAR, SHAMEBOY

The most memorable characteristic of sensitive briar is that its leaflets fold up when they are touched. The prickly branches, growing three or more feet long, trail along the ground and display beautiful pink flower clusters among the delicate bipinnate leaves.

Mimosa helps to stabilize sandy soil. The pink blooms appear in April to June and may bloom into the fall.

Strophostyles helvula (Phaseolus helvolus)

SAND BEAN, TRAILING WILD BEAN, ANNUAL WILD BEAN

Trailing along a sandy lakeside or in the disturbed soil of a country road, the small purplish-pink flowers of sand bean are an exquisite surprise with their distinctive inward-bending keel. The bean pod splits open when dry, releasing the fuzzy seeds, which readily spread volunteer seedlings to stabilize the soil. This annual vine bears trifoliolate (with three leaflets) compound leaves.

Sand bean is attractive to butterflies and bees, and its seeds provide food for game birds. Native Americans boiled and mashed the edible roots. Iroquois are said to have rubbed the crushed leaves on their skin to remove warts and to relieve the discomfort of poison ivy rash. A boiled decoction of the bean has been used to treat typhoid.

A view of the leaf arrangement is found on p. 225.

Tephrosia lindheimeri

HOARY PEA, SHOESTRING PEA, LINDHEIMER'S TEPHROSIA

This native vine of the central and southern portion of the state beautifies the landscape with white-bordered pinnate leaves and showy pink flowers. Hoary pea is recommended as a ground cover and will grow well in shaded areas. It is also tolerant to both cold and drought and has the added benefits of attracting butterflies and of enriching the soil with nitrogen.

Trifolium pratense
RED CLOVER, PEAVINE CLOVER, COWGRASS

This valuable clover species was introduced to England in the mid-1600s, and the British in turn brought it to America, where it is now found naturalized in the higher-rainfall areas of Texas.

Red clover is widely cultivated as a nutritious forage plant for livestock, comparable in value to alfalfa. Humans also appreciate the flavor and consume the dried flowers as a beverage, which has a mild sedative or relaxing effect.

Red clover has been considered a remedy for everything from athlete's foot to cancer. It is commercially important as a source of isoflavones and phytoestrogens and is sold as a dietary supplement for relief of hot flashes and other unpleasant side effects of menopause. Red clover tea is touted by herbal healers to help ease a respiratory illness accompanied by a nagging cough. It is also used to treat water retention, constipation, skin irritation, and ulcers.

GENTIANACEAE (Gentian Family)

Centaurium beyrichii
MOUNTAIN PINK, ROCK CENTAURY, QUININE-WEED

Mountain pinks, and their close relative rosita, are members of the gentian family, which was named for King Gentius, an ancient ruler of the Balkans who used gentians to cure a fever among his soldiers.

Rosita (shown on the Exploring Further pages) is slightly taller, has fewer and larger blooms, and has lanceolate (lance-shaped) leaves, whereas the mountain pink has very narrow, linear leaves. Rosita's stamens are in the corkscrew shape of a honey dipper, and the stigma is bifurcated, or split in two at the end. It is found alongside streams and stock tanks, while its cousin, the mountain pink, prefers dry, rocky soil.

Mountain Pink, Rock Centaury, Quinine-Weed

These plants contain phenolic acid and are useful laxatives, antacids, fever reducers, and appetite stimulators. One of mountain pink's specific uses has been in the treatment of malaria, hence the name "quinine-weed." It also kills parasitic worms. Ancient Egyptians used other gentian species to treat kidney stones. Bitter juice from the root was the key ingredient in a soft drink, known as "Beverage Moxie Nerve Food," that was popular in the late 1800s.

Both species are toxic to deer and livestock, causing animals digestive distress and even ulceration if the grazing is prolonged. Kidney and liver damage may also occur and result in death. Deer avoid rosita, making it useful in landscaping where deer resistance is important.

A close-up of rosita is shown on p. 225.

KRAMERIACEAE (Ratany Family)

Krameria lanceolata

TRAILING RATANY, PRAIRIE SANDBUR, KRAMERIA

The ratany family has only one genus with about twenty-five species and is found from South America to the southern United States. Many ratany species are low shrubs, of which three are found in the Trans-Pecos region. *Krameria lanceolata* differs from other ratany species in that its stems trail along the ground—hence, one of its common names, "trailing ratany." It is found throughout Texas and grows in a variety of soil types.

Another of the common names, "prairie sandbur," refers to its spiny fruits. A quick glance at the young fruit may lead one to think that this is the goathead of the caltrop family, but closer inspection will reveal soft spines and a covering of downy fuzz; at maturity, these spines prick with needlelike pain. The fruits are often inhabited by moth larvae, which eat the seed embryo. Prairie sandburs are partial parasites; that is, they can photosynthesize to produce their own carbohydrates, but they also send out root fibers to raid the food supply of other plants.

The maroonish-pink flower is an intriguing study in intricate detail. Two of the petals fuse into a green fan shape that looks like a cobra rearing its head. Instead of typical nectar, ratany flowers produce an oil that is attractive to pollinators, particularly bees of the genus *Centris*, who feed it to their young. These insects have specially adapted legs to harvest the oil.

Tea made from the roots can be used to heal sore throats or oral infections such as mouth sores. A tea brewed from the leaves also soothes stomach ulcers or other problems in the lower digestive tract. Prairie sandbur has astringent properties that constrict blood vessels, thus making the plant useful in stopping bleeding from wounds.

A view of the prickly seedpod is found on p. 225.

LAMIACEAE (Mint Family)

Monarda clinopodioides
BASIL BEEBALM

Springtime in East and Central Texas finds basil beebalm in glorious bloom—a treat for the eyes as well as for nectar-loving pollinators. The remarkable arrangement of basil beebalm's flowers resembles a shish kebab or a wedding cake. Its potential for adding a unique vertical element to a landscape makes it welcome in any wildflower garden.

All members of genus *Monarda* contain thymol, an antiseptic and a parasitic worm killer. Native Americans made soothing salves, or balms, from *Monarda*. Beebalm tea can be used to bring down fever as well as to relax an insomniac. A word of warning: excessively strong tea may induce vomiting.

Salvia greggii
AUTUMN SAGE

The genus name *Salvia* comes from a Latin word meaning "to be saved" and refers to the ancient belief that growing sage in the garden prevented illness from entering the home. Perhaps there is a bit of truth to that

old myth, because *Salvia* does help maintain good health in those who are wise in the ways of herbs. The herbal tea brewed from the aromatic leaves may be effective as a decongestant or as an aid to drying up milk production when a mother is trying to wean her nursing infant. The tea soothes sore throats and upset stomachs and is used as a treatment for diarrhea.

Autumn sage is a drought-tolerant plant and is now a popular ornamental in native landscapes.

MALVACEAE (Mallow Family)

Callirhoe involucrata var. *involucrata*
WINECUP, POPPY MALLOW

According to a legend from India, winecups came into being when a servant was performing a dance to cheer up the beloved king, who had fallen ill. The servant placed a goblet filled with red wine in the open palm of his hand and danced on the lawn until he finally stumbled from exhaustion. Everywhere the wine spilled, a flower grew up, having the shape and color of the goblet of wine.

The winecup has a root that is similar in flavor to a parsnip or a sweet potato and was prized as a cooked vegetable by Indians of the Great Plains. The leaves are edible, and, like most mallows, they contain a dense, slippery mucilage that is useful for thickening soups.

Winecup flowers' stunning good looks make them popular as ornamentals. The plants grow best in well-drained soil, in full sun or partial shade. Unfortunately, the tender young plants are relished by slugs, which will destroy them if left unchecked.

Winecup, Poppy Mallow

Like many members of the mallow family, the winecup has unusual-shaped seedpods that resemble cheese "rounds," as shown in the photo on p. 226.

Malva sylvestris

HIGH MALLOW, ZEBRA HOLLYHOCK, CHEESEFLOWER

The high mallow, native to Eurasia, has escaped cultivation because of its free-seeding habit. This versatile plant produces prolific quantities of purplish-pink flowers, even in low-moisture conditions.

The young leaves are tasty and nutritious when boiled; the seedpods, resembling cheese rounds, are also edible. At one time high mallow was one of China's most important green vegetables.

Europeans once used high mallow to make garlands and door decorations to celebrate May Day. Early Romans made a drink from it as a hangover remedy. Pliny, the first-century Roman naturalist, regarded a daily teaspoonful of high mallow drink as an essential preventative for all diseases. European herbalists made a strong infusion of leaves and flowers to treat kidney stones, and both Europeans and North American

settlers used it as a poultice to reduce pain and swelling caused by insect bites. It also soothes sore throats, as well as teething pain in babies.

Along with its other relatives, high mallow contains a slimy mucilage that provides emollient (soothing) and demulcent (softening) properties for skin abrasions and irritations. It also contains red and blue pigments that make it useful as an indicator of acidity or alkalinity.

Pavonia lasiopetala

ROCK ROSE, PAVONIA MALLOW

Rock rose is sometimes called the pavonia mallow after the Spanish botanist José Pavón (1750–1844), who studied South American vegetation. This shrubby perennial resembles a wild rose when viewed from a distance, but up close it clearly looks like its relative, the hibiscus. Fine hairs at the base of the petals have earned it the species name *lasiopetala*, meaning "shaggy petals."

Because it is drought tolerant, rock rose makes an outstanding addition to a xeriscape. Blooming prolifically, it adds a strong statement of color and has the added attraction of luring butterflies to the garden. Rock rose plants are perennial but have a short useful life of only three to four years. The plants reproduce readily from seed.

Native Americans have used the bark of rock rose for cordage and rope.

A view of the mature fruit and seeds is found on p. 226.

NYCTAGINACEAE (Four O'clock Family)

Abronia angustifolia

NARROW-LEAF SAND VERBENA, PURPLE SAND VERBENA

Sand verbena is not a verbena, but the flowers can fool someone into thinking it is. The other part of its name is true—it does live in sandy regions of West Texas and the Trans-Pecos across to Arizona. The fruit looks like a jester's hat, or the "twinkle-twinkle-little-star" toy for children.

Over most of its range it grows as an annual, but it is a perennial in the White Sands area of New Mexico, where it is able to tap into a reliable water supply.

At dusk, sand verbena releases a strong fragrance, drawing moths and other insects.

A view of the quaint fruit is found on p. 226.

Allionia incarnata

TRAILING FOUR O'CLOCK, PINK WINDMILLS, TRAILING ALLIONIA

This genus is named for the eighteenth-century Italian botanist and medical doctor Carlos Allioni, who was a friend of the well-known taxonomist Carolus Linnaeus.

Trailing four o'clock is a bit of a deceiver. For one thing, what appears to be one flower is actually three. In this little cluster the three strongly bilateral blooms hold their "petals" all to one side so that when viewed together, the triplets will appear as one blossom. Furthermore, those "petals" are actually brightly colored sepals. True petals can be distinguished from sepals by the number of vascular bundles supplying each structure: petals and anthers have one each, while sepals and leaves have three.

Notice that the leaves, which are opposite each other at each node, are of two different sizes; this special trait is characteristic of the four o'clock

Trailing Four O'clock, Pink Windmills, Trailing Allionia

family. Early morning is the best time to see the delicate purplish-pink flowers, but they remain on display throughout most of the day, unlike those of other members of this family.

Native Americans used trailing four o'clocks to relieve stomachache. This practice would be very dangerous without knowledge of dosage size, because trailing four o'clock contains potent toxins that are halluci-nogenic. In fact, some tribes used this mind-altering trait when seeking visions as part of religious ceremonies.

ONAGRACEAE (Evening Primrose Family)

Gaura spp.

GAURA, BEE BLOSSOM, WAVING BUTTERFLY

Gaura comes from a Greek word meaning "superb," a fitting name for this handsome wildflower. Another picturesque name, "waving butterfly," was conferred because its four petals resemble fluttering butterfly wings, and the long anthers, the butterfly's antennae.

There are a number of *Gaura* species, all difficult to differentiate because they hybridize easily. *Gaura* can be distinguished from other members of the evening primrose family by the tall column of unopened buds above the blooms. The flowers open in the morning, and the petals wilt by afternoon. Colors range from white to pink or red. Often a blossom that is white in the morning will be a darker color later in the day. Each flower has a distinctive four-part stigma, which is typical of this genus.

Bee blossom is now commercially available in many plant nurseries, and homeowners can enjoy its good looks as well as the butterflies and bees that are attracted to its sweet-smelling flowers.

Gaura also has practical uses as a natural remedy for rheumatism and as a treatment for burns and inflammation. The Lakota people are said to have used it as a "horse catcher" by rubbing the plant on their hands before going out to round up their horses.

Oenothera canescens

SPOTTED PRIMROSE, BEAK-POD EVENING PRIMROSE

Spotted primrose inhabits the Panhandle region, thriving in the clay soil of drying playa lakes. Unlike the other more flamboyant evening primroses, the diminutive spotted primrose can easily escape one's notice, but the cute pink flower, with its well-exserted stigma and its squatty, winged seedpod, is well worth the effort of keeping one's eyes glued to the ground.

The name *canescens* refers to the white, hairy covering of the leaves.

Photos of other evening primrose species are found on pp. 95, 96, and in the next entry.

Oenothera speciosa

PINK SHOWY PRIMROSE, PINK LADIES

The masses of cotton candy–colored blossoms that grace Texas roadsides every spring are known as pink showy primroses, which the eminent wildflower enthusiast Lady Bird Johnson called one of her favorites.

Although the pink petals seem very delicate, the showy primrose is actually a hardy, drought-resistant plant that does very well in landscapes as long as the soil is well drained. Sometimes it does a little too well and becomes an invader, but the bees, moths, and hummingbirds won't complain. Pink showy primrose seeds easily and also spreads prolifically by sending out rhizomes, or underground stems, to generate new floral progeny.

Tender young leaves may add a pleasant flavor to a salad or be cooked as a green vegetable. A good-quality yellow dye can be extracted from this plant.

Photos of other evening primrose species are found on pp. 95, 96, and 212.

PAPAVERACEAE (Poppy Family)

Argemone sanguinea
ROSE PRICKLY POPPY, RED POPPY, SPINY PRICKLEPOPPY

The genus name *Argemone* is a Greek word meaning "cataract of the eye," referring to the belief that the bitter yellow, or even orange, sap could

be used to treat eye problems. (Note: One should never use this remedy since the sap contains protein-digesting enzymes and can permanently damage the eye.) The species name *sanguinea* means "blood colored," an obvious reference to the pigment of the petals.

Rose prickly poppies are among the showiest of wildflowers, with their large blooms (often more than three inches across) atop stems that may be up to four feet tall. The petals are usually a shade of pink or lavender but may be white. Close examination of the flowers reveals numerous stamens that have yellow, or even red, filaments supporting the anthers. The central pistil is topped with an eye-catching purple stigma to receive the abundant pollen.

Rose prickly poppies bloom from February through May, especially in disturbed soil in the southern and western parts of the state.

For photos of other colors of prickly poppy, please see pp. 98 and 268.

PEDALIACEAE (Sesame Family)

Proboscidea louisianica
DEVIL'S CLAW, UNICORN PLANT

The flowers of the devil's claw are beautiful, but the sprawling plant's stickiness and odor are less easily appreciated. The green seedpods have a curved, hornlike extension, hence the name "unicorn plant." When the pod dries, it earns its other common name, "devil's claw," because the extension splits in two and the resulting "claws" catch easily in animal fur or on their feet.

When sliced and fried, the sticky, young, green seedpods look and taste very similar to okra. They also make tasty pickles. The small, black seeds can be extracted from the dried seedpod by pulling on both "horns" and breaking the pod in half. The dark seeds can be eaten much like sun-

flower seeds. The seed coats are cracked off, revealing the edible, nutlike interior, which is high in both protein (20–35 percent) and oil (35–45 percent). The seeds' high unsaturated fatty acid content gives them potential value either for culinary use or as a drying oil in varnishes and paints.

The Hopi people believed that the curved arms of the dried pods would attract lightning and therefore bring rain, so the plants were never weeded from their cultivated fields. Some people in the Mexican American culture have a negative perception of devil's claw, believing that stepping on a seedpod brings bad luck.

There is good evidence that the Pueblo people practiced selective breeding of devil's claw for white seeds and longer pods. Fibers from the dried black pods are still used by Native Americans in making decorations on baskets.

Views of the immature and mature fruit are shown on p. 226.

POLEMONIACEAE (Phlox Family)

Phlox drummondii

DRUMMOND PHLOX

Drummond phlox's exquisite flowers became famous throughout the world as a result of Thomas Drummond's botanical explorations to the New World in the early 1800s.

The word *phlox* is believed to come from the Greek word for "flame," evoking the bright colors of this species. Several subspecies of phlox grow wild throughout Texas, some in shades of red or pink and others in purple, lavender, or white, and occasionally, even a yellow one. The five-petaled blooms have a ring or a star in the center, which is one of the hallmarks of members of the phlox family.

Phlox is a fast-growing, prolific-blooming annual that begins its floral show as early as February in milder parts of Texas and continues through the early summer. The plants look particularly striking in large groups.

In the romantic days of old, there was an elaborate code between friends and lovers in which various flowers were symbolic of particular sentiments. In this so-called language of flowers, phlox signified an agreement or consensus. If a certain question was asked and a phlox was sent by way of reply, then the answer was "yes."

POLYGONACEAE (Buckwheat Family)

Polygonum bicorne (Persicaria bicornis)

PINK SMARTWEED

Pink smartweed is recognized by its cherry-red stems and short, more-or-less upright flowering heads. Pale smartweed, shown on the Exploring Further pages at the end of this section, is less vivid, and its long flowering heads droop gracefully. These species are common in and around playa lakes of West Texas and the Panhandle, in ditches, or in other low-lying areas.

Smartweed takes its name from the stinging sensation felt in the mouth when the strongly acidic juice meets soft tissue, causing it to "smart." The genus name *Polygonum* means "many knees or joints," in reference to the swollen leaf nodes that are depicted below and in the family photo on p. 15. Encircling each node and petiole of its attendant leaf is a papery sheath, the ocrea, which is a modified stipule.

Smartweeds have been used to treat a variety of ailments. A tea made from the plant tops has been used in treating epilepsy, and a leaf tea was given to women to control bleeding after childbirth. An old remedy for hemorrhoids advocated the use of smartweed leaves to clean the affected area.

The blossoms, which are borne in elongated flower groupings called racemes, are an important source of nectar for bees, and the seeds are valued by birds and small mammals.

Many species in this genus have value as food, and archaeologists have discovered smartweed seeds in ancient campsites, indicating its long history of use. American Indians ground the tiny black smartweed seeds for flour. The starchy root is also edible.

The young, tender leaves may be eaten raw or cooked, but because the mature foliage contains significant amounts of oxalic acid, the older leaves should be cooked in several changes of water. Oxalic acid has the immediate effect of numbing the mouth and making the tongue swell and, on a longer-term basis, may aggravate conditions such as gout, kidney stones, and arthritis. Tannins are present in a number of smartweed species, and both yellow and purple dyes can be extracted as well. Unfortunately, students have had a notable lack of success in using smartweed to help them pass exams.

Polygonum bicorne, *pink smartweed; close-up of flowering head*

Polygonum bicorne, *pink smartweed; close-up of red stems and "swollen" joints*

Close-ups of the cherry-red stems and pink flowers of *P. bicorne* are shown here. Nodding smartweed, a paler, droopier relative of pink smartweed, may be seen on p. 226.

PORTULACACEAE (Purslane Family)

Portulaca pilosa
SHAGGY PORTULACA, PINK PURSLANE, KISS ME QUICK

The origin of shaggy portulaca's common name is readily evident with a glance at the leaf axils, which are covered in long white hairs. Even its specific name, *pilosa*, means "covered with soft hair." The purplish-pink petals complement the bright yellow stamens. Unfortunately, the flowers last only a short time each day, wilting as the day heats up.

In addition to its being a desirable ground cover in a rock garden, shaggy portulaca's fingerlike succulent leaves are edible and nutritious.

In Brazil, portulaca is a traditional medication for pain relief, and it has also been used as a sedative as well as a treatment for diarrhea.

SAPINDACEAE (Soapberry Family)

Ungnadia speciosa
MEXICAN BUCKEYE

Despite the name, this species is not a true buckeye, but the seeds closely resemble those of its namesake: shiny and black with a single white spot. People carry seeds of both species for good luck.

Mexican buckeye, a large shrub or small tree that reaches up to thirty feet tall, may be found on rocky hillsides and in canyons. Lovely pink blossoms cover it in the spring, making it easy to recommend as a remarkable ornamental feature for home or park gardens.

It contains toxic glycosides and alkaloids in young vegetative growth as well as in the mature leaves and seeds. These poisons affect mammals and birds, and even honeybees feeding on the nectar will be killed. Beehives should be removed from the vicinity of Mexican buckeye plants during the flowering period.

The toxic seeds have reportedly been used by Indians to poison arrow tips. Crushed seeds thrown into water are said to stun fish, making for easy (but unethical and illegal) fishing.

A view of the seedpod is found on p. 226.

SCROPHULARIACEAE (Snapdragon Family)

Penstemon ambiguus
PINK PLAINS PENSTEMON, GILIA BEARDTONGUE, PINK PLAINS BEARDTONGUE, SAND PENSTEMON

Unusually for members of the genus *Penstemon*, this species has quite small flowers, but that does not detract at all from the plants' showiness because the stems of the plant are massed together in a mounded semi-snowball shape. The flower itself is full of graceful curves—a bit uncertain and "ambiguous" about their direction.

Pink plains penstemon puts on its best show early in the spring and may flush with flowers throughout the growing season if water is available. It is extremely drought and heat tolerant. Furthermore, butterflies and moths are frequent visitors to the flowers. These attributes make it a highly desirable member of a native landscape.

Pink Plains Penstemon, Gilia Beardtongue, Pink Plains Beardtongue, Sand Penstemon

For more information about *Penstemon* characteristics and medicinal properties, see the next entry.

Penstemon guadalupensis

GUADALUPE BEARDTONGUE, WHITE PENSTEMON

The Guadalupe River is an important feature of Central Texas, where Guadalupe beardtongue makes its home. Like many other *Penstemon* species, it has showy flowers, ranging from pink to white, that make it a valuable addition to a native landscape. A close examination of the mouth of *Penstemon* flowers will reveal the five stamens for which the genus received its Latin name. Extending from the mouth of each bloom is a staminode, or sterile stamen, bristling with fine hairs and making one think of a bearded tongue, hence the common name.

Bees are a major pollinator of beardtongue flowers, and the capacious design of the "tube" allows the insects plenty of wiggle room in their quest for nectar. The dark markings on the floor of the corolla show them the way to the sugar formed in the nectaries at the base of the stamens. In the process, the insects brush against the anthers and receive a powdering of pollen, which they will carry to inadvertently pollinate the next beardtongue flower. Moths, butterflies, and hummingbirds are also frequent visitors and serve as pollinators as well.

The Navajos used *Penstemon* to treat arrow or gunshot wounds, based on the ability of these plants to inhibit inflammation and to promote new tissue growth. They also used it to treat skin wounds of their animals. Native Americans mixed the crushed stems with beeswax or oil to treat chapped lips and irritated skin. Other *Penstemon* species were used as laxatives.

For a field view of Guadalupe beardtongue, see p. 227.

Verbascum blattaria

MOTH MULLEIN

Verbascum blattaria is too pretty to have a name with such a disgusting meaning: *blatta* is Latin for "cockroach," which this plant is said to repel. The common name, "moth mullein," comes from the efficacy of the foliage in driving away the small moths that eat holes in clothing when it is packed away for the winter. Some people also seem to think the downy stamens of the flower resemble the hairy antennae of a moth. An extract from moth mullein, which is possibly the same chemical that repels roaches and moths, is effective in killing mosquito larvae.

Moth Mullein

Moth mullein displays a slender spire of five-petaled blooms atop four- to six-foot-tall plants. The dense purple fuzz covering the red stamens is an intriguing detail in the center of each flower.

This native of Eurasia grows wild in the eastern part of Texas, where it provides food for wildlife; finches especially relish the seeds. Moth mullein has become a popular ornamental as well.

TAMARICACEAE (Tamarisk Family)

Tamarix ramosissima

SALT CEDAR, TAMARISK

Salt cedar is not a true cedar, but it was named for the small, scalelike leaves that resemble those of its namesake. This shrubby tree has wispy inflorescences of tiny pink flowers (rather than the cones produced by cedars).

Native to southern Europe, salt cedar was introduced in the United States to stabilize sandy soil and to serve as a windbreak. It has naturalized along the banks of salty streams and serves as an indicator of salinity. Since it tolerates the salty environment of the seashore, it is a useful landscape plant around beach houses, helping to prevent erosion of the unstable sandy land. In many areas, however, salt cedar has become invasive and has made itself so much at home in its adopted land that eradication programs are being implemented along the Pecos River, the Rio Grande, and other watercourses in the western part of Texas.

Although livestock will not consume salt cedar, it is useful for humans in an unexpected way. When insects damage the stems, the plant exudes a sweet, sticky sap known as "manna"; this sugar is collected and used in confectionery.

A close-up of the inflorescence is found on p. 227.

Spines and other prickly structures help protect plants and their fruits, as well as help in dispersing the seeds. This section serendipitously has several examples of these botanical strategies. Other fruiting structures shown here cling or take to the air.

Asclepias speciosa, *showy milkweed; tubercled, or "spiny," seedpods (follicles) that are unlike the smooth seedpods of other milkweed species, p. 186.*

Carduus nutans, *musk thistle; immature flower head, p. 187.*

Centaurea americana, *basket flower; note the basketlike network of phyllaries beneath the young flower heads, p. 188.*

Cirsium horridulum, *horrible thistle; close-up of flowering head, p. 189.*

Echinacea angustifolia, *purple coneflower; field appearance showing the drooping petals typical of this species, p. 190.*

Echinacea angustifolia, *purple coneflower; mature flower head stiff enough to use as a comb! p. 190.*

Palafoxia sphacelata, *sand palafoxia; note stalked glands on stem and phyllaries under the flower head, p. 191.*

Chilopsis linearis, *desert willow; field appearance, p. 192.*

Echinocactus texensis, *horse crippler; note red fruit and daggerlike appearance of spines, p. 193.*

Opuntia imbricata, *cholla; field view, p. 194.*

Strophostyles helvula, *sand bean; vines bearing trifoliolate leaves (compound leaves with three leaflets), p. 199.*

Centaurium calycosum, *rosita; larger than its relative, mountain pink, and inhabits a wetter environment, with stamens that are corkscrewed like a honey dipper, p. 201.*

Krameria lanceolata, *trailing ratany; seedpod—as wicked as it looks! p. 203.*

Callirhoe involucrata *var.* involucrata, *winecup; immature seedpod, looking like a cheese round, with seeds—white structures that will mature to dark brown, p. 205.*

Pavonia lasiopetala, *rock rose; mature seedpod—brown eggs in a basket! p. 208.*

Proboscidea louisianica, *unicorn plant; immature fruit in the "unicorn" stage, p. 214.*

Abronia angustifolia, *narrow-leaf sand verbena; immature fruit, p. 208.*

Proboscidea louisianica, *devil's claw; mature fruit in the "devil's claw" stage, p. 214.*

Polygonum lapathifolium, *nodding smartweed; note drooping inflorescences, p. 216.*

Penstemon guadalupensis, *Guadalupe beardtongue; field appearance, p. 220.*

Tamarix ramosissima, *salt cedar; close-up of flowers, p. 222.*

Ungnadia speciosa, *Mexican buckeye; mature fruit with shiny black seeds with white "eye," p. 218.*

White Flowers

AGAVACEAE (Agave Family)

Dasylirion texanum

SOTOL, DESERT SPOON, DESERT CANDLE

This stately sentinel accents the hills of West and Southwest Texas with flowering spikes reaching five to twenty feet in height. Sotol is dioecious, with male and female flowers residing on separate plants. Shown here is a close view of the flowers of a female specimen. The one-seeded fruit is triangular with three wings. The leaves sheathe the stalk with spoon-like bases, and their assymetric shape makes an interesting focal point in dried floral arrangements.

Archaeologists have evidence that Native Americans from Mexico and Southwest Texas used sotol for food and fiber for thousands of years. In a land of few trees, the tall stalks were used for fuel and for construction, and they are still used for that purpose in some areas. The fibrous leaves can be used to make sandals, mats, hats, ropes, and baskets.

The immature flower stalks have a high sugar content and are valued as cattle feed. This sugar content was also used to produce a colorless

alcoholic beverage called "sotol" by first roasting and then fermenting the "heart." People have also roasted or boiled the bases of the leaves for food. In times of drought, ranchers burn off the leaves and split open the heart of the emerging flowers for cattle to eat.

Sotol's flowering spires make objects of vertical interest in a xeriscape design.

Close-ups of the male flowers and of the leaves, along with the field view, are found on p. 282.

Yucca glauca (*Y. angustifolia*)

YUCCA, BEARGRASS, SOAPWEED

In the late 1500s the British doctor and horticulturist John Gerarde mistakenly thought the plant to which he was giving the name *Yucca* was cassava, or "yuca," as it is called in the Caribbean and in Central America. They are not at all related, but the confusion persists even today.

Yucca and the *Pronuba* moth have a reciprocally beneficial relationship, with interdependent life cycles. Each species of *Yucca* relies solely on a particular species of the small white *Pronuba* moth for pollination of its creamy white flowers. The female moths have specialized mouthparts to carry out the pollination and, immediately afterward, insert their eggs into the yucca's ovary. The moth caterpillars emerging from the eggs feed upon some of the seeds but leave enough untouched to ensure the perpetuation of the yucca generations. Highly favored flowers host an entire bevy of female moths, who make themselves comfortable in the flower throughout the day.

Yucca flowers, which are high in vitamin C, are nutritious food for cattle or people. They make a mild-flavored garnish to a salad, or they can be sautéed with onions and tomatoes for a tasty vegetable dish. The young green fruits, shown on the next page, remind some people of squash and are good to eat either chopped or stir-fried.

The roots are toxic, however, and should not be eaten, unless one is seeking the benefit of an energetic laxative, because they contain the glycoside saponin. Saponin has a more valuable use as a cleansing agent of another sort: chopped roots mixed with water yield a sudsy solution for shampoo and laundry and for cleaning baskets and wool rugs.

Yucca shampoo is thought to enrich the appearance of one's hair, and application of root material was believed to cure baldness. Perhaps this was related to yucca's value in killing head lice. Ceremonial shampoos were important aspects of certain Native American weddings and funerals. Some Native American legends say that dreaming of washing one's hair in yucca roots foretells death.

The roots of yucca contain significant quantities of phytosterols, which are used by the pharmaceutical industry to manufacture hormones and other steroid products. Tea from the root is used as a remedy for arthritis and for urinary and prostate inflammation. Lakota women mixed the roots of yucca with prickly pear cactus roots to help in difficult childbirth, calling the remedy "medicine for not giving birth."

Before cotton was introduced from South America, yucca fiber was used extensively for manufacturing rope or twine for house construction and for making bowstrings, fishing nets, baskets, sandals, and clothing. Yucca spines were an essential component of a snakebite kit; the spines were jabbed into the area around the bite to stimulate bleeding, and, it was hoped, discharge of the toxin. American Indians exploited the subduing effect of the smell of burning yucca roots to make their horses easier to capture.

Below are photos showing female *Pronuba* moths in a yucca flower, as well as yucca fruits in the tender, young stage when they are edible. Note the wilted petals still clinging to the maturing ovaries.

A yucca with dried, mature seedpods may be seen on pp. 282–283. Three other species of yucca are also shown there.

Female Pronuba *moths in* Yucca *flower*

Yucca glauca, *yucca; immature seedpods*

ASCLEPIADACEAE (Milkweed Family)

Asclepias asperula

ANTELOPE HORNS, SILKWEED, INMORTAL

The genus *Asclepius* is named in honor of the Greek god of healing, and its importance in medicine is reflected in its extensive use in Native American and Spanish herbal remedies. Many species have similar chemical properties: the milky sap, from which the group derives its common name, and the roots are the principal bearers of various glycosides that act on involuntary muscles, such as the heart and uterus.

As a rule, all milkweeds should be considered poisonous, but the narrow-leaved, whorled species such as *A. engelmanniana* (shown on the next page), contain glycosides that act on the nervous system and are considered to be the most dangerous.

Milkweed flowers have an alien anatomy so unlike that of "normal" flowers that botanists use a unique vocabulary to describe the structures. The stigma resembles a five-cornered helicopter landing pad with the anthers clamped vertically on the sides. Within the anthers, the pollen grains are linked in a saddlebag arrangement.

The mature seedpods (called follicles) split open when dry, releasing the downy-tufted (comose) seeds. Native Americans collected the seed fluff to spin into fiber for clothing and to make string or fine cord for fastening feathers to prayer sticks. During World War II, the downy fibers served as fillers for life vests.

According to folk medicine lore, the milky sap is good for the complexion, and it has been specifically used to fade freckles and remove warts. The sap of most milkweeds is antifungal and antibacterial and has been used to heal chronic skin sores. Some Native American tribes used this plant to relieve bronchial and nasal congestion: they chewed the latex, which would loosen phlegm and open the breathing passages. This practice seems a bit risky, however, considering the presence of cardiac glycosides in the fluid.

The powdered roots have been recommended as a treatment for headache and fever. A natural method of clearing a stuffy nose is to snort powdered milkweed root, which causes violent sneezing. The diuretic properties of the liquid extracted from boiled roots made it useful for treating mild kidney complaints and for relieving breast engorgement in nursing mothers. The roots of some species were chewed by the Blackfoot people to treat sore throat and painful gums, and the Lakotas used its pulverized roots as a gentle treatment for children with diarrhea. Native American women used the root as a contraceptive. However, if this birth control method failed, they used it during labor to reduce pain and help lessen the contractions following childbirth.

Given its poisonous properties, it may seem surprising that some parts of milkweed become edible if they are boiled in several changes of water. Many Native American peoples cooked the young flowers and the young shoots as vegetables and used the immature seedpods, which contain an enzyme known as asclepain, to tenderize buffalo meat.

The plant's use by the Lakota, Hopi, and other Native Americans to stimulate milk flow in nursing mothers is a typical example of the

Asclepias oenotheroides, *hierba de zizotes; flowers and dehisced follicles*

Asclepias engelmanniana, *narrow-leaved milkweed; whole plant*

Doctrine of Signatures, which is an ancient belief that the characteristics of a plant would give a clue to its medical properties—in this case, that the milky sap would promote milk production.

Milkweeds are sought out by monarch butterflies as a nursery for their eggs so their hatchlings can incorporate the sap's poisonous glycosides into their bodies as they feed on the foliage. The caterpillars become so bitter tasting that birds learn to avoid them. Even the adult butterflies have enough toxin remaining in them to give protection from predatory birds. Although the plant is dangerous to grazing animals at all stages of growth, after a killing frost it is no longer toxic.

In the photos on the previous page, compare the long, narrow leaves of *A. engelmanniana* with the short, wide foliage of *A. oenotheroides*. Note the seedpods of both species, one of which has dehisced to release seeds.

Views of seedpods in different stages of maturity are found on p. 283. For additional information on milkweeds, see pp. 38, 150, and 186.

ANACARDIACEAE (Sumac Family)

Rhus microphylla
LITTLELEAF SUMAC, WINGED SUMAC, AGRITO

Littleleaf sumac, a shrub of the desert grasslands, grows in association with mesquite, juniper, or creosote bush throughout the western part of the state. It tolerates low rainfall conditions of three to sixteen inches and grows at elevations up to six thousand feet. Littleleaf sumac flowers early,

Rhus microphylla, *littleleaf sumac; bush in bloom.*

between March and May, with the prolific white blossoms appearing before the leaves.

Sumac's hardy growth habit, delicate foliage, and bold red berries make it a candidate for xeriscaping, either as an individual shrub or, with judicious pruning, as an attractive hedge of several sumacs. Mule deer browse it, birds love the fruit, and caterpillars find the leaves tasty and may defoliate it at certain times of the year.

All parts of several sumac species have been used by Native Americans and are presently used by modern-day herbalists for a range of needs, from medicines to drinks and dyes. Sumacs contain significant amounts of tannins, giving them astringent properties for constricting soft tissues. The powdered roots, for example, were used to stop bleeding. The leaves were brewed to make a tea to alleviate the discomforts of the common cold, and some American Indians also chewed the berries to treat colds and flu as well as to relieve toothache. Antibiotic compounds have subsequently been identified in the berries. A soothing emollient for the lips or other chapped body parts can be made by combining powdered sumac leaves with Vaseline or glycerine.

Sumac's tart fruits make a refreshing homemade "lemonade." Its roots also yield a yellow dye.

The profusion of flowers in a sumac tree is shown above. The colorful mature fruits, which are covered with hairs, are shown on p. 283.

ASTERACEAE (Sunflower Family)

Achillea millefolium

YARROW, MILFOIL, THOUSAND LEAF

The genus name *Achillea* comes to us from tales of the Trojan Wars as told by Homer. When the warrior hero Achilles received his famous wound in the heel, a Greek god appeared with yarrow leaves to stanch the flow of blood. Yarrow was believed to be especially effective for wounds caused by iron weapons and remained an important tool for battlefield surgeons through the nineteenth century.

Achillea is a native of Europe and Asia and is now naturalized in North America. Its frilly, finely divided leaves give it a fernlike appearance that is recognized both by its species name and its common name "milfoil," or "thousand leaves." For the adventurous diner, the fresh young leaves can spice up a salad with a peppery flavor.

Yarrow was highly rated as a medicinal herb in its native Europe, so it is not surprising that early settlers in North America brought it with them. Its anti-inflammatory properties make it valuable as a treatment for skin wounds and eczema, for digestive and urinary disorders, and for female reproductive disturbances. A tea made from the leaves has been used to lower fever, reduce sleeplessness, treat earaches, and decrease menstrual bleeding. Nonetheless, its chemical portfolio contains compounds that tend to cause dermatitis and therefore reduce *Achillea*'s attractiveness for topical use.

A number of important ecological studies with *A. millefolium* have shown that the appearance of plants of the same species may vary considerably when grown under different climatic conditions. Sometimes plants of identical genetic makeup may be erroneously classified as separate species when, in fact, the plants are simply responding to their individual environments with a different growth habit.

Aphanostephus skirrhobasis

ARKANSAS LAZY DAISY

Arkansas lazy daisy's sophisticated-sounding Greek name, *Aphanostephus*, simply means "inconspicuous wreath" and refers to the small, pretty flowers, suitable to crown a fairy's head, or to the small tuft of hairs on the seed, for which a microscope is needed in order to fully appreciate their glory.

The Arkansas lazy daisy is an annual with soft, hairy leaves. When its ray petals are clasped together in the evenings and mornings, the undersides reveal a delicate pink hue, but when the flowers get around to opening toward noon (yes, they are lazy), a new persona is exposed: the topsides of the petals are a bright white, making the whole group of flowers look like a group of little ballerinas dancing in their white dresses.

Erigeron spp.
FLEABANE DAISY, BRUISEWORT

There are about fifteen species of fleabane daisy in Texas, all of them quite similar in appearance. The many narrow, bright white ray flowers (or "petals") are attached in two rows; the petals of some species show pinkish-lavender on the underside.

Its genus name, *Erigeron*, means "growing old quickly," referring to how rapidly the flowers fade and are replaced by seeds wearing long white fluffy hair, or pappus, on their upper end (think of Albert Einstein's hair).

Plants in this genus are cute enough to welcome into a garden, but before succumbing to the temptation, be forewarned that fleabane daisy seeds prolifically, and its abundance may soon be a source of regret.

The fleabane daisy was named for the belief that its unpleasant smoke could drive away fleas, and early American settlers would hang bunches of this plant in their homes to keep these pests away. Native Americans crushed the leaves and rubbed them into their dogs' fur to rid them of fleas and other hangers-on.

Some Native Americans powdered the disk flowers to use as snuff, inducing a sneeze, which relieved the symptoms of a stuffy head cold. Fleabane tea was used to treat a wide-ranging host of ailments, including mouth sores, urinary problems, rheumatism, digestive upsets, and tonsillitis. This plant has also been called "bruisewort" because it was used as a poultice to treat swellings and bruises.

The Sioux used the flowers in a concoction of buffalo brains and gall as a tanning agent for buffalo hides.

Eupatorium serotinum

LATE-FLOWERING BONESET, FALSE BONESET, LATE BONESET

The genus name *Eupatorium* originates from Mithridates Eupator, who was king of Pontus about 100 BC. Legend has it that he discovered that plants in this group could be used to treat poison victims. The *serotinum* part of the name comes from the Latin word *serum*, meaning "late." The "late-flowering" appellation is in reference to its September through November bloom period. As the name suggests, boneset was utilized to help heal broken bones.

Late-flowering boneset colonizes moist, disturbed areas with an underground network of rhizomes. Plants display groups of white flowers atop two- to six-foot-tall stalks.

Although many insects find late-flowering boneset to be a delectable feast of pollen and nectar, mammalian herbivores avoid the bitter foliage. *Eupatorium* may therefore become common in overgrazed areas due to preferential consumption of more palatable plants.

A tea brewed from the leaves has been employed as a febrifuge (fever reducer) and as a diuretic. The Houma Indians used it to treat typhoid fever.

Evax prolifera

RABBIT-TOBACCO, CARAS DE LOS MUERTOS, COTTON ROSE

Had Lewis Carroll placed a special order for plants to populate his *Alice in Wonderland* tales, among them would have been *E. prolifera*. Its common name "rabbit-tobacco" reflects the belief that it was a favorite food of rabbits, who supposedly used the plant as a "chaw" of tobacco.

The plant itself is tiny and fuzzy, and the flower heads are a further study in miniature, featuring grotesque little skull-like faces (*caras de los muertos*) that peer expectantly up at their audience. Indeed, the flower anatomy is so strange and obscure that most of the recognizable features of the sunflower family are absent. There are no normal phyllaries (bracts) around the flowers and no ray flowers; the disk flowers, such as they are (the little skull faces), are fertile only around the edges of the flower head. With such an apparent lack of vigorous reproductive facilities, it is all the more surprising that rabbit-tobacco can proliferate and become an invasive weed in the dry, alkaline soils that it prefers.

Melampodium leucanthum
PLAINS BLACKFOOT, BLACKFOOT DAISY

The genus name *Melampodium* means "black foot," referring to the dark, foot-shaped bract under the ray flowers. The name *leucanthum* indicates that plains blackfoot has white flowers.

These fresh-looking daisies grow well in a variety of soil types, including disturbed areas. The growth habit is low and bouquetlike, and they make beautiful borders and focal points in a native plant landscape. Their allure has not been lost on the flower seed and bedding plant industries, which have made several varieties available on a commercial scale.

Navajos prepare an eyewash from blackfoot daisy.

Parthenium confertum

FALSE RAGWEED, LYRELEAF PARTHENIUM, GRAY'S FEVERFEW

The name *Parthenium* indicates its medicinal use in "women's problems"; the name is derived from the Greek word *parthenion*, referring to a chorus for girls or to a "virgin's bower."

Lyreleaf parthenium is a plant of disturbed areas and prefers the dry, calcareous soils of West Texas, although it tolerates a range of soil types, including a dry corner of a home garden. It reseeds easily but does not aggressively displace neighboring plants. The plant serves as browse for deer.

Other *Parthenium* species provide quininelike compounds for treatment of fevers; the root is used to treat urinary infections, and the leaves are used as a poultice for burns. Industrial uses have been found for guayule, *P. argentatum*, which was developed as a source of latex for tires in World War II when supplies from Southeast Asia were threatened.

A view of the whole plant is found on p. 284.

Pinaropappus roseus

ROCK-LETTUCE, PINK DANDELION

Rock-lettuce flowers resemble those of dandelions (except the color is white, sometimes with rosy undertones), making their appearance in the spring and continuing through the summer. These attractive, drought-tolerant perennials are about six inches tall and are suitable for native plant landscaping in rocky, limestone areas.

Traditional healers in Mexico extract juices from the boiled plant to apply to injuries as a general antiseptic. They also recommend drinking rock-lettuce tea to someone suffering from indigestion or constipation. The stems exude a white sap that has traditionally been used to treat toothaches. Research has revealed that *Pinaropappus* contains several flavonoid antioxidants.

CACTACEAE (Cactus Family)

Opuntia leptocaulis

CHRISTMAS CACTUS, TASAJILLO, JUMPING CACTUS

Like a tree decorated for the holidays, the Christmas cactus is ornamented with bright red fruits that may remain on the plant through the next flowering season, giving it a year-round festive look. These fruits are deemed delectable by birds and deer. Humans also find the fruit good to

Christmas Cactus, Tasajillo, Jumping Cactus

eat, but the tiny morsel is hardly worth the effort of removing spines and seeds.

Joints break off easily and attach themselves to unwary passersby, hence the name "jumping cactus."

The small, pale blooms appear in July and August and attract many pollinators. Bird droppings disperse the seeds, accounting for the growth of Christmas cactus under areas where birds perch, such as along fence lines.

The eye-catching fruits and slender stems are shown on p. 284.

CAPPARIDACEAE (Caper Family)

Cleome serrulata

CLAMMYWEED, ROCKY MOUNTAIN BEE PLANT, SPIDERFLOWER

When Lewis and Clark trekked across America, they encountered an incredible number of new (to them) species of wild plants. The lacy white to pink clammyweed was one of these and was described by the German botanist Frederick Traugott Pursh in his book on the expedition's floral discoveries, *Flora Americae Septentrionalis* (1814).

Of course, the plant was already well known to the native inhabitants of North America, who had been using it for centuries. Despite its unpleasant odor, the plant is edible. They ground the seeds to make flour and ate the leaves, stems, and flowers like boiled spinach, changing the water several times to reduce the bitterness.

Native Americans put the crushed leaves on insect bites or on sore eyes and made a tea to bring down fever and to soothe an upset stomach. The leaves, boiled with corn to improve the flavor, were given to people suffering from intestinal problems. A traditional treatment for anemia is to drink a tea made from the flower boiled together with a rusty nail.

By boiling the whole plant, Native Americans extracted a black dye used for decorating pottery.

Possibly the most memorable, and ironic, use for this smelly plant was as a deodorant for both the body and the shoes—or moccasins, as the case may be.

CHENOPODIACEAE (Goosefoot Family)

Salsola tragus (S. iberica)

TUMBLEWEED, RUSSIAN THISTLE

Introduced from Russia in the late 1800s, tumbleweeds have spread rapidly across the southwestern United States, thanks to their "specialized" seed dispersal method, in which the entire mature plant breaks loose and rolls across the countryside scattering seeds.

Although tumbleweeds would never win a beauty contest, the tiny, delicate, and generally overlooked flowers along the stems are worth the effort of inspection.

The tender seedlings make a tasty cooked green vegetable—one that will never be in short supply in West Texas. Boil them until they turn a

Tumbleweed, Russian Thistle

dark green, add a sprinkle of salt, a little sugar, and some butter to gain a new appreciation for this traditional enemy of farmers.

In general, tumbleweeds are a safe source of nutrition for grazing animals; however, if the plants are grown in areas such as cattle pens or fertilized areas, the tumbleweeds may concentrate nitrates, causing harm or even death to the animals.

Views of the seedling and the whole plant are found on p. 284.

CONVOLVULACEAE (Morning-Glory Family)

Convolvulus equitans
BINDWEED, TEXAS BINDWEED

Bindweed's flowers are lovely to look at, but it is a detested invader of gardens and croplands, where its twining habit strangles other plants and deprives them of light, water, and nutrients. Its unsavory reputation is somewhat counterbalanced by a captivating legend that has the flowers serving as the tiny dresses of fairy children.

Convolvulus equitans comes in two color forms: one is pure white, and the other, the one no doubt preferred by Texas A&M fans, is white with a maroon center.

A view of the fruits is found on p. 284.

Cuscuta cuspidata
DODDER, LOVE-VINE, DEVIL'S GUTS

If you've seen stringy yellow threads decorating plants along the roadside or, heaven forbid, in your garden, you probably felt a little disgusted at the sight, and rightly so. The plant is dodder, a parasite that, surprisingly enough, is in the morning-glory family. Dodder begins life as a seed, and it germinates and grows until it attaches to a plant by means of haustoria, or modified roots. Once the haustoria are connected, the actual root then withers, and dodder carries on as a rapidly growing, perpetual dependent.

Because of the destructive effect on its hosts, dodder has a repertoire of unsavory names, including hellweed, beggarweed, strangle tare, witches' shoelaces, tangle gut, and devil's guts. By contrast, it is also called "angel hair" because it shines with a golden color in the sunlight.

Dodder, Love-Vine, Devil's Guts

Despite its negative reputation, dodder does have some ardent adherents. The name "love-vine" arose from the practice of hopeful young men twirling around three times with a bit of vine and throwing it over their shoulder. If his sweetheart loved him, the vine would grow, and not surprisingly, the vine almost always showed she did.

There are about 170 species of dodder: some prefer a specific host, but others are quite eclectic. Taking not only nutrients but also many other chemicals from their host, dodder may have varying medicinal properties that are linked to those of the host plant.

All dodder species are bitter and act as a purgative. Both seeds and vegetative material are useful for treating urinary tract infections. Cherokees crushed the orange "threads" to make a poultice for bruises, while other tribes used dodder as a contraceptive. People with lung illnesses were made to sit in a bath with crushed dodder.

Dodder is a source of a yellow or orange dye, which the Pawnees extracted for coloring feathers.

Control of dodder requires persistence, as no single method works effectively. Dodder doesn't grow in winter, and it doesn't grow on grasses and other monocots, so ornamental grasses, lilies, or elephant ears are recommended for infected flower beds.

Merremia dissecta (Ipomoea sinuata)

ALAMO VINE, NOYAU VINE, CORREHUELA DE LAS DOCE

Alamo vine differs from most of its morning-glory relatives in that it doesn't bloom in the morning but waits until noon to unfurl its twisted bud. The large white blossom that it reveals has a reddish-purple throat in which the uniquely curled anthers reside. The flower woos pollinators all afternoon and then closes as the sun sets. Once pollination is accomplished, the white corolla withers and falls away to reveal a pepper-shaped seedpod. The palmate leaves typically have five to seven lobes and are almost as interesting as the flowers.

The Alamo vine makes an attractive ornamental, especially in its natural home in the southern and central parts of Texas. It blooms from May to November. When making wildflower selections for your garden, remember the Alamo (vine)!

Views of the fruits and leaves are found on p. 284.

For additional information on morning-glory species, see pp. 139 and 196.

CRUCIFERAE or Brassicaceae (Mustard Family)

Capsella bursa-pastoris
SHEPHERD'S PURSE

Like all members of the mustard family, shepherd's purse forms two-compartment seedpods between the flowering head and the leaves. This plant was named for the resemblance of its seedpods to the bags that shepherds carried. Although the pods are romantically heart shaped, the actual shepherds' purses were made from ram scrotums.

The young leaves of shepherd's purse, especially before the flower stalk appears, make a tasty and nutritious addition to a salad, or they can be served as a cooked vegetable seasoned with mushrooms and thyme. The leaves are high in vitamins A, B_1, B_2, C, and K, and they provide iron as well. The seeds and roots are also edible but less flavorful.

Shepherd's purse is used to treat diarrhea as well as to relieve bladder infections, and drinking the tea is said to lower blood pressure. The astringent properties of shepherd's purse allow it to be used for constricting blood vessels and for contracting the uterine muscles, which has made it a friend of women in childbirth or with menstrual problems. Pregnant women should never use shepherd's purse because its muscle-contracting properties could be disastrous for the baby. People with sensitive skin

should be cautious in using this plant as a skin astringent, since blistering could result.

The seeds are said to be toxic to mosquito larvae and may be put in standing water to help control these pests.

Dimorphocarpa wislizenii

SPECTACLE POD

An attentive observer of a vast white colony of spectacle pods will not only note their stately beauty but also focus on the myriad pairs of old-fashioned spectacles returning the gaze. The snowy flowers are borne at the tops of the eight- to twenty-inch-tall stems, but the really intriguing parts of the plant are the seedpods, which are found just below the cluster of blooms. All members of the mustard family have two-chambered pods, but none are so amusingly cute as those found on this plant.

Although it's easy to see how the plant got its common name, the scientific name seems more mysterious until one realizes that *Dimorphocarpa* means "two-shaped fruit," in reference to the two-chambered seedpod. The species name *wislizenii* is in honor of Friedrich Wislizenus (1810–89), who came to America from Germany and collected plants throughout the West.

Native Americans enjoyed eating the entire plant, fresh or dry. The seeds, which are hot and spicy, were eaten parched or fried, or they were dried and stored for winter. Meat was given a piquant flavor when cooked with this tasty herb.

The Hopi people employ spectacle pods to treat wounds.

Lepidium oblongum

PEPPERGRASS

The several species of peppergrass have small, scalelike seedpods between the young flowers and the leaves. In fact, the genus name *Lepidium* means "little scale."

The common name "peppergrass" comes from the fact that the seedpods and leaves are hot and spicy and may be added to food to impart a peppery flavoring. When eaten raw, the plant has a taste resembling that of broccoli, which is also in this family. Peppergrass is tasty in a salad, giving a sharp-flavored counterpoint to the mild-flavored lettuce.

The Great Lakes Indians crushed *Lepidium* plants as a treatment for poison ivy and other skin rashes. Native Americans discovered they could chew the leaves to relieve a headache and could brew a tea to treat kidney problems. In caring for their animals, they mixed the crushed plant with lime to kill maggots infesting any wounds.

Tan, yellow, and green dyes can be extracted from *Lepidium*.

EUPHORBIACEAE (Spurge Family)

Cnidoscolus texanus

BULL NETTLE, TREAD-SOFTLY

Bull nettle may look pretty and smell sweet, but don't be fooled. It has a nasty, prickly side to it that could make one very sorry to have made its intimate acquaintance.

Bull nettle is a large, bushy perennial plant growing to about three feet wide and two to three feet high. The fragrant white flowers are showy and enticing, but beware of the plant's prolific covering of bristly hairs and caustic irritants! A simple touch may cause a painful rash, particularly because the hairs tend to break off in the skin. A person who inadvertently makes contact with the stinging hairs can find relief in the application of a weak solution of ammonia. Windex is a good choice, but out in the countryside the only available ammonia may be in urine. Desperate times call for desperate measures.

However, very young nettles are edible before they develop their spiny adult character. Each pod produces three seeds, which the intrepid Native Americans also used as food. The seeds are said to be appetizing, with a nutty flavor.

Both plants and animals have methods of making sure their children move away from home, and bull nettle sends its offspring out into the world by catapulting the seeds out of the ripe pod.

A view of the whole plant is found on p. 285.

Euphorbia marginata
SNOW-ON-THE-MOUNTAIN

Both snow-on-the-mountain and its close relative snow-on-the-prairie (shown in the Exploring Further pages) are whimsical reminders of a snowfall, hence their common names. The lovely white "petals" are actually modified leaves, or "bracts."

Snow-on-the-mountain grows mostly in the Panhandle and Central Texas, and it has shorter leaves than the very similar but longer-leafed snow-on-the-prairie, which is found in the eastern third of the state.

Their height (four to five feet) and the crisp contrast between the green and white colors allow both of these species to serve as focal points in a native plant landscape. Snow-on-the-prairie thrives best in heavy clay or loamy soil, while snow-on-the-mountain prefers calcareous soil. Neither species will be bothered by the depredations of deer and rabbits.

The broken stems exude a white sap containing a poisonous chemical known as euphorbium, which may harm the skin or eyes. Ranchers have been known to use the caustic latex to brand cattle. The juices aggressively irritate the mouth and the intestinal tract of cattle that eat the plant, but sheep and goats are unaffected.

In early times many people believed that plants gave clues to their medicinal uses by some visible characteristic. This belief, called the Doctrine of Signatures, led the Zuñi Pueblo people and the Lakotas to use the milky-sapped snow-on-the-mountain to encourage milk flow in nursing mothers. Small amounts of the toxic sap have also been used as an intestinal purgative.

Snow-on-the-prairie (*E. bicolor*) is shown on p. 285.

FABACEAE or Leguminosae (Bean or Pea Family)

Astragalus racemosus

ALKALI MILKVETCH, CREAM MILKVETCH, RACEME MILKVETCH

Alkali milkvetch's creamy white flowers are attached to the stems in clusters called "racemes," leading to both its scientific name and one of its common names. The pinnately compound leaves resemble those of the common mesquite tree, a fellow member of the legume family. The pendant bean pods are triangular in cross section, a trait that helps set apart alkali milkvetch from its close relatives.

Like other *Astragalus* species, it is poisonous due to the presence of alkaloids and accumulation of selenium.

A close-up of the leaves and fruits is found on p. 285.

For a comparison of *Astragalus* with *Oxytropis*, please see p. 160.

For other photos of *Astragalus*, see p. 156.

Dalea candida (*Petalostemum candidum*)
WHITE PRAIRIE CLOVER

Prairie clover's genus name, *Dalea*, honors the English botanist Samuel Dale (1659–1739). Inhabiting caliche or rocky soils from Mexico to the prairies of southern Canada, prairie clover's lacy white flowers would also look right at home in a wildflower garden. This perennial produces several erect to spreading stems that may reach three feet in height.

Many Native Americans ate the raw roots as a snack and made tea from the dried leaves. The tea causes slight constipation, making it useful for treating diarrhea.

A view of the whole plant is found on p. 285.

Desmanthus illinoensis

ILLINOIS BUNDLE FLOWER, PRAIRIE MIMOSA

Despite its name, Illinois bundle flower is not confined to that state but is found throughout Texas in all but the southernmost parts. It has an intriguing, tightly twisted cluster of seedpods, and the leaves have a lacy appearance. Like some others of its leguminous relatives in the mimosa tribe, the bundle flower's leaves are sensitive and will fold together when touched. This perennial may grow up to three feet tall.

The seeds serve as food for quail and other birds, and livestock find this nutritious, high-protein plant to be very palatable. It serves as an indicator of good range conditions; if found in a pasture, it means that the number of livestock is being well managed.

Illinois bundle flower's past use as a treatment for the eye disease trachoma was definitely not for the fainthearted or sensible. Five seeds were placed in each affected eye at night, and the eye was then washed with water the next morning.

Native American children used the dried beans for rattles, and today the clusters of curved seedpods are used in dried flower arrangements.

A view of the mature seedpods is found on p. 285.

Eysenhardtia texana

TEXAS KIDNEYWOOD, VARA DULCE

Eysenhardtia grows in the calcareous soil of the Trans-Pecos area, the Edwards Plateau, and South Texas. It was named in honor of the German botanist Karl Wilhelm Eysenhardt.

Texas kidneywood is an irregularly shaped shrub of eight feet or more in height that is relished by goats and white-tailed deer as nutritious browse. The leaves give off a strong, rather unpleasant odor when crushed, but perhaps that spices up the diet of animals in the way that picante sauce does for people.

Texas kidneywood, with its drought tolerance and lengthy bloom period from April right through the summer into fall, deserves a place in a native plant landscape. The spikes of small white flowers give off a vanilla-like fragrance that honeybees adore and that attracts the Arizona skipper butterfly, for which it serves as a host plant.

Texas kidneywood has been found to contain both antibacterial and antifungal agents, explaining its success as a traditional medicine for treating kidney and bladder infections.

Trifolium repens

WHITE CLOVER, LADINO CLOVER, DUTCH CLOVER

This low-growing clover features clusters of white flowers that are some-times tinged with a bit of pink. The compound leaves generally have three leaflets, but sometimes a lucky four-leaf clover is encountered, as poet Ella Higginson (1861–1940) commemorates in this little verse:

> One leaf is for hope, and one is for faith,
> And one is for love, you know,
> And God put another in for luck,
> If you search, you will find where they grow.

White clover's original home is in the eastern Mediterranean region, from whence it spread throughout Europe and reached the Americas along with the European settlers. Because of its symbiotic relationship with the nitrogen-fixing bacterium *Rhizobium trifolii*, it enriches the soil wherever it grows and has been an important factor in maintaining fertil-ity in pastures and farmland.

White clover is a favorite for honeybees, sheep, and cattle. Highly nutritious as well as palatable, white clover is cultivated as a forage crop, but it has also spread into lawns as an invasive weed.

LAMIACEAE (Mint Family)

Monarda pectinata var. *punctata* (*M. punctata*)
SPOTTED BEEBALM

The genus *Monarda* is named in honor of Dr. Nicholas Monardes, a Spanish botanist and physician who wrote a book in 1596 describing the plants of the New World. Spotted beebalm's varietal name, *punctata*, means "dot" or "point," referring to the purple spots of color on the creamy background of the corolla.

Spotted beebalm is recommended for a native landscape because of its long flowering period (from May until July, or longer) and its ability to draw swallowtail butterflies, bees, and hummingbirds to the garden.

Like all members of the genus *Monarda*, spotted beebalm has medicinal properties that have a long history of use. It contains thymol, which is effective in eliminating parasitic worms. This chemical is also used to make cough syrups. Tea brewed from the plant is used to bring down fever, but the tea should not be too strong, as it might cause nausea. Native Americans used this plant to make a skin-soothing balm, hence its common name.

Teucrium laciniatum

CUT-LEAF GERMANDER

Cut-leaf germander thrives in a variety of habitats across the western two-thirds of Texas. Its contrasting fresh white corollas and green foliage and its low-growing habit make this a bonny border plant for a wild-flower garden. It grows well under both arid and moist conditions.

Germanders as a group have been used to make alcoholic beverages or teas that are digestive aids. They may also serve as gout treatments and as diuretics to relieve water retention. The juice from the boiled plant has been used as a wash for wounds and as an anti-inflammatory agent for irritated gums.

People harboring parasitic worms have given the unwanted guests the old heave-ho by downing germander tea. However, overindulgence in germander products may result in liver damage.

A green dye may be extracted from the leaves.

LILIACEAE (Lily Family)

Allium drummondii

WILD ONION

With flower colors ranging from white to pink or purple, wild onions are a winsome wildflower, but many people think of them only as weeds that invade lawns. Wild onions have an underground bulb, although much smaller than that of their domesticated cousins, which may be eaten cooked or raw. The plants contain significant quantities of vitamins A and C, and the bulb contains inulin, a type of starch made of fructose that is not readily digested and acts as a laxative and, even more noticeably, as a flatulent.

The Blackfoot Indians boiled wild onions to make a tea for treating coughs and relieving nausea and vomiting. They charred the bulbs and inhaled the smoke to treat colds and sinus headaches; they also used the plants to make a wash for irritated eyes and to treat ear infections. The juice from crushed bulbs can soothe bee and wasp stings. Wild onion eliminates intestinal worms. It also has a positive "heart health" effect by lowering triglyceride levels.

Animals that eat large quantities of this plant may suffer poisoning from N-propyl disulfide. Cattle, horses, and even cats are especially susceptible to this red blood cell–destroying chemical. Animals who have had prolonged exposure to it may become weak and jaundiced, with

dark red or brown urine. Diagnosis is made easy by the animal's "onion breath."

Native Americans extracted a brownish-orange dye from wild onions for coloring the yarn used to weave blankets. They also rubbed the juice from the crushed bulbs on their skin to repel insects (and possibly a few friends).

In the Exploring Further section (p. 285), one view shows the floral parts in multiples of three, a characteristic typical of monocots. The adjacent photo depicts mature ovaries splitting to release black seeds.

Nothoscordum bivalve

CROW POISON, FALSE GARLIC

The genus name *Nothoscordum* comes from the Latin words meaning "false garlic." It superficially resembles garlic and onions, both members of the genus *Allium*, but it is not a close relative, nor is it edible. The flowers of crow poison differ from those of wild onion chiefly by the presence of yellow in the center. If in doubt, just smell the root. If it smells like an onion, it is an onion and is safe to eat. False garlic will simply smell like the dirt in which it grows.

Crow poison has the typical six tepals (three sepals and three petals that are identical in appearance) of the lily family. An observant person will be able to tell the sepals from the petals by gently closing the flower and noticing that three sepals will be on the outside and three petals will be on the inside.

Crow poison grows well in sandy soils, particularly in disturbed sites. It spreads prolifically by seeds and underground bulbs and may form extensive colonies.

LOASACEAE (Stickleaf Family)

Mentzelia decapetala

TENPETAL BLAZINGSTAR, TENPETAL MENTZELIA

The heady-scented flowers of *M. decapetala* have large petals and resemble a lotus flower; they open as the sun sets, and hordes of bees arrive for frenzied feeding. *Mentzelia nuda* flowers, shown on the following page, have more numerous and narrow petals; they open in the morning and remain open through the day.

When not in bloom, *Mentzelia* would not draw anyone's attention, but when it blooms, its full glory is displayed. The curious-looking fruit resembles a candle with a wick, which is actually the remnant of the style, or pipelike structure, through which the pollen tube travels on its way to the ovary to deliver the male gametes.

The most distinguishing feature is the leaf, which bears a multitude of minute barbed hairs that cling tenaciously to fabric and fur. Stickleaf species inhabit the same terrain of limestone hills and canyons as sheep, much to the dismay of ranchers, because the leaves are virtually impossible to remove from the wool and therefore reduce its market value.

The Dakota people extracted and boiled down a thick, yellow juice from the stems to use as an externally applied medicine for fever. The Zuñi Pueblo people made a suppository from the powdered root to relieve constipation. It was a favorite of the Cheyennes, who harvested the root for treating fevers, ear infections, and arthritis. Tea made from the roots of *Mentzelia*, together with *Echinacea*, was used to care for the unfortunate victims of diseases brought by Europeans, including measles, mumps, and smallpox. Seeds were roasted and powdered to sprinkle on the skin eruptions caused by smallpox to keep the sores from becoming scarred and pitted.

Mentzelia nuda, sand lily; close-up of flowers.

To assuage thirst, Great Plains Indians chewed the roots of both sand lily and *Echinacea*. (See *Echinacea* on p. 190.)

The seedpods of *M. decapetala* are shown on p. 286.

MALVACEAE (Mallow Family)

Callirhoe involucrata var. *lineariloba*
WHITE WINECUP

Callirhoe involucrata var. *lineariloba* flowers vary in color from pure white to the same magenta color as its better-known cousin, *C. involucrata* var. *involucrata* (see p. 205). Like its cousin, it is drought tolerant yet will thrive under more hospitable conditions, sending its trailing stems with masses of leaves to a distance of four feet or more from the main stem. The cup-shaped flowers extend above the foliage, showing off to their best advantage.

White winecups attract many butterflies, and the gray hairstreak uses the foliage as a nursery for its caterpillars. Deer also love this plant.

White Winecup

Those who practice traditional medicine advocate using a tea made from the root or inhaling the smoke from burning the root to alleviate internal pain.

Malva neglecta

DWARF MALLOW, CHEESE PLANT, LOW MALLOW

The mild-tasting, nutritious leaves and young shoots of dwarf mallow may be eaten raw in a salad or cooked as a green vegetable. The mucilage, which is typical of all mallows, will thicken a soup. The roots contain so much of this slick, slimy substance that it has been put to use as a substitute for egg whites in making meringue.

A tea made from dwarf mallow leaves has been used to soothe a sore throat, alleviate the pain of bladder infections, and calm an upset digestive system. It also acts as a laxative gentle enough for children. Among Native Americans the tea is traditionally given to mothers to aid in

childbirth, and afterward it is used to wash the newborn baby. The soft, downy leaves, whether fresh or dried, are useful as a poultice for painful inflammations. Among its valuable constituents are phytosterols, which help to reduce blood cholesterol levels.

Native Americans have used the root of dwarf mallow as a toothbrush. They also extracted dyes ranging in color from light tan to yellow or green.

NOLINACEAE (Nolina Family)

Nolina texana

BUNCH-GRASS, BEARGRASS, SACAHUISTE

Nolina texana, or bunch-grass, is not in the grass family, as its name and appearance might suggest, but its family designation has gone through more than one change. Originally placed in the lily family, it was later moved to the agave family, only finally to be given its own group, Nolinaceae.

This stemless perennial has a profusion of long, wiry leaves that may grow up to four feet long, draping gracefully over rocky ledges in its native home in Central Texas to the Rio Grande plains and the Trans-Pecos area.

The one- to two-foot-long clusters of tiny white blooms make their appearance from March to July, nestling within the crown of leaves. Bunch-grass is dioecious, forming male and female flowers on separate plants, but there will be some blossoms containing both sexes on each plant.

Bunch-grass contains saponins, which are toxic for sheep and goats, and to a lesser degree, cattle. Early symptoms of *Nolina* poisoning are itchy dermatitis and photosensitivity. Animals will lose their appetite

and become jaundiced as the liver and kidneys are damaged. The toxin appears to be mainly in the buds, flowers, and fruits, so access to pastures where *Nolina* is present should be limited during the bloom period.

Nolina's saponins were a valuable source of soap for Native Americans. The saponins are also the active ingredient for treating arthritis and rheumatism, as well as vascular problems and lung ailments. The flowering stalk is edible if boiled or roasted, although it is said to taste soapy.

Native Americans used the long leaves as a source for strong, flexible fibers to make mats, cords, snares, and whips. Some tribes customarily used bunch-grass mats for baby cradles, while others used the mats to cover their dead. Apaches placed layers of the leaves on the floors of their tepees, and others used the foliage to thatch the roofs of their wickiups. Navajos still use *Nolina* as a source of dye for their blankets.

Nolina makes an excellent choice for easy-care landscaping due to the plant's resistance to drought, heat, and cold. It is particularly attractive in settings where the leaves can extend or drape to their full length. Bunchgrass has an additional advantage of being deer and rabbit resistant.

PAPAVERACEAE (Poppy Family)

Argemone polyanthemos
WHITE PRICKLY POPPY, CRESTED PRICKLY POPPY

The genus name comes from the Greek word *argemone*, which means "cataract of the eye," referring to the belief that the bitter yellow or orange sap could be used to treat eye problems. (Note: One should never actually attempt to use this in the eye since the sap contains protein-digesting enzymes and can permanently damage the eye.) The species name *polyanthemos* means "many flowers."

White prickly poppies are large, showy plants that grow in sandy soil,

waste places, or overgrazed pastures. They may reach three feet in height and display huge flowers that are about four inches in diameter.

For photos of other colors of prickly poppy and for additional information on these species, see pp. 98 and 213.

PLANTAGINACEAE (Plantain Family)

Plantago helleri

HELLER'S PLANTAIN

The genus name *Plantago* means "like the sole of a foot," in reference to the shape of the leaf. Native Americans called plantain "white man's foot" because it seemed to grow wherever Europeans had been.

The flowers, seeds, and roots are used for their laxative effects and for treating hemorrhoids. Indeed, the seed and seed coat of *P. psyllium* or *P. ovata* are the main ingredients in the commercial laxative products Metamucil and Perdiem. *Plantago* seed coats contain up to a third of their weight as hydrocolloids, which absorb water and help move things along, and also have a cholesterol-lowering effect. In fact, the U.S. Food and Drug Administration approved an official health claim in 1998 that allows food manufacturers to say on the package label that foods containing psyllium fiber may lower the risk of cardiovascular disease.

Protein-digesting enzymes in the fresh leaves and roots of plantain constrict blood vessels to reduce bleeding, and Native Americans and settlers used the leaves to bind wounds and to prevent infection. Plantain juice was used to soothe ulcers, irritated intestines, and inflamed hemorrhoids. The tea served as a treatment for urinary tract infections, diarrhea, and chronic lung problems. Chewed or crushed leaves were used as a poultice for insect stings, poison ivy rash, wounds, and burns. The dried, powdered roots were used for toothache pain.

Plantago is high in vitamins A and C, and the leaves may be eaten raw in salads. Leaves have also been smoked as a ceremonial tobacco.

POLYGALACEAE (Milkwort Family)

Polygala alba
WHITE MILKWORT, SNAKEROOT

The genus name *Polygala* is derived from the Greek for "much milk." Milkwort ("milk herb") has long been believed to increase the milk flow in nursing mothers, and farmers have added it to the diet of dairy cows for that purpose as well.

White milkwort resembles the Seneca snakeroot, *P. senega*. The former is found in the southern and western prairies of the United States, while the Seneca snakeroot is found in the Canadian prairies and eastern United States. The roots of both species are highly valued for their medicinal properties and were bartered among the various Native American tribes during the 1700s and 1800s.

The Indians in the Great Lakes region used milkwort primarily as a decongestant and expectorant for relieving flu and cold symptoms, as a treatment for insect bites and stings, and as a sedative for nervous conditions. They also used it to treat heart disease and to alleviate ear infections. It was believed to be a treatment for snakebite and thus was called "snakeroot," but there is little indication of its actual efficacy in that regard.

Its medical value was recognized early on by European doctors settling in North America, and by 1740 it had made the journey across the Atlantic and become established in English gardens. It is still popular in Europe as an ingredient in cough remedies, and several thousand tons are exported annually from Canada to Europe, Japan, and the United States for use in herbal medicines. Traditional Chinese medicine also makes use of snakeroot. However, as is the case with many herbal remedies, no controlled trials have been conducted to verify snakeroot's effectiveness.

Milkwort contains methyl salicylate, better known as oil of wintergreen, which is used in liniments, but its important active ingredients are a host of saponins, which may cause severe negative gastrointestinal responses—vomiting and diarrhea—if consumed carelessly.

POLYGONACEAE (Buckwheat Family)

Eriogonum multiflorum
BUCKWHEAT, HEARTSEPAL BUCKWHEAT

Buckwheat's genus name means "woolly knees," referring to the swollen and hairy leaf nodes of members of the Polygonaceae ("many knees") family.

The numerous flowers (*multiflorum*) of heartsepal buckwheat are in clusters of pastel pink, cream, or white. The plant's grayish-white appearance is due to a fine covering of white hairs.

Buckwheat species inhabit the rangelands and canyon breaks of the Panhandle and southwestern Texas.

Several species of *Eriogonum* were used by the Navajo, Paiute, Hopi, and Shoshone peoples for medicinal purposes. The pain-killing properties of these plants were used as an analgesic for rheumatic pain. Buckwheat also provided treatment for tuberculosis, bladder trouble, coughs, and hemorrhaging. The flowers relieved water retention, soothed irritated or inflamed membranes, and made an effective gargle for a sore throat.

Among the Hopis, several buckwheat species are associated with "women's medicine" and are used to relieve menstrual pain as well as back pain suffered by pregnant women. The plant was also used to assist in childbirth and as a bath for newborn babies.

Although buckwheat is not considered a major food source for humans, the Hopis boiled buckwheat leaves and combined them with cornmeal to bake a type of bread.

Dried mature buckwheat plants make a striking focal point in a dried flower arrangement.

A related species, grassland buckwheat (*E. alatum* var. *glabriusculum*), shown on p. 286, has fruits that are winged (*alatum*).

RANUNCULACEAE (Buttercup Family)

Anemone berlandieri
TENPETAL ANEMONE, WINDFLOWER

Although the name would indicate that there are ten "petals," there are actually none at all. What you see are showy sepals, which range in color from white to blue or purple. The individual pistils, or female parts, form a conelike structure in the center of the flower.

According to mythology, the Greek god of the wind, Anemos, sends windflowers, or tenpetal anemones, to herald his coming in the blustery month of March. Ancient Romans believed that picking the first anemones in the spring was good luck and would prevent illness throughout the year. By contrast, Egyptians thought of this flower as the emblem of sickness, and the Chinese used it in funerals, calling it the "flower of death."

Tenpetal anemone is toxic and should never be eaten, but small amounts of it have been used to treat measles, headaches, diarrhea, and

coughs. Herbalists recommend it for calming nerves. Native Americans found that wounds and boils healed more quickly when the crushed fresh plant was applied. Recent research has shown tenpetal anemone to have antimicrobial properties that help prevent infections.

Delphinium carolinianum
PRAIRIE LARKSPUR, ESPUELA DEL CABALLERO

The genus name *Delphinium* comes from the Latin word for "dolphin," which the profile of the larkspur flower resembles, if one uses a little imagination. The name "larkspur" was conferred because its modified calyx was believed to look like the leg spur of the lark. Its other common name, "espuela del caballero," is based on its similarity to cowboy spurs.

The flowers of prairie larkspur are usually a grayish- to bluish-white. The flowers contain both male and female parts, but self-pollination is inhibited because the stamens (male structures) mature before the pistils (female structures).

All parts of this plant are poisonous and should never be eaten. The alkaloid concentration is highest in the spring after flowering and declines as the plant grows older. Cattle are especially susceptible to the toxins, which cause paralysis of the respiratory system, loss of motor control, and sudden death.

Native Americans found that although larkspurs could not be used as food, the plants were useful in other ways. Larkspur yields a pale orange dye. The Pueblo people used larkspur petals as part of some ceremonies, and the Kiowas used the seeds in peyote rattles. The plant's poisons make an effective treatment for lice as well as scabies, but their toxicity is so great that they should be used only if it is a great deal more trouble to find a pharmacy than a handful of larkspur.

ROSACEAE (Rose Family)

Fallugia paradoxa
APACHE-PLUME, PONIL

Apache-plume has specially decorated seed heads in which the numerous seeds each trail a long, feathery appendage. These purplish clusters were thought to resemble the war bonnet of an Apache chief, hence the common name. Apache-plume's genus name, *Fallugia*, is in honor of Abbot V. Fallugi, a nineteenth-century botanist from Italy.

The bark and flowers were used by Native Americans for much the same purposes as we use aspirin: to bring down a fever and to relieve pain. The Indians brewed a tea from the root to calm a cough and consumed the tender stems for indigestion. A leaf tea was used as a tonic to thicken the hair. They made a poultice of the crushed flowers to bring down leg swelling, both in their own limbs and those of their treasured horses. When they wanted to relieve an aching joint, they mixed the leaves with tobacco and applied the poultice to the painful area. The indigenous peoples discovered that eating Apache-plume petals helped relieve gas pain. They had so much faith in the powers of this plant that it was believed that drinking it would magically repel hexes from witches.

Besides being a cure-all, Apache-plume was also used in fence making. The strong, flexible roots were used like cords to tie fence posts together.

SAPINDACEAE (Soapberry Family)

Sapindus saponaria var. *drummondii*

SOAPBERRY, JABONCILLO, PALO BLANCO

Soapberry grows throughout Texas along watercourses and at the edge of wooded areas. From March to July it puts on a pleasing show of clusters of small white flowers that will develop into wrinkled orange berries. It also makes an exceptional shade tree for home landscaping, growing from thirty to forty-five feet tall.

As one might surmise from the common name, the berries are used for soap. The berry juice lathers when mixed with water. The fresh fruits may be used, or they may be crushed to extract the liquid to save for washday. This soap is considered to be of excellent quality for washing delicates and for shampooing hair.

A tea from the leaves is an analgesic for acute arthritis pain, and it also reduces inflammation. Washing in the berry juice may halt the spread of some skin fungi.

Native Americans used soapberry to catch fish by stunning them. The saponins in the fruits enter the fish's bloodstream directly through the gills, stupefying them so that they float to the surface for easy collection; however, this practice is now considered unethical and is illegal. This poison is believed to have no effect on the taste of the fish.

A view of the fruit is found on p. 286.

SCROPHULARIACEAE (Snapdragon Family)

Penstemon albidus

WHITE BEARDTONGUE, RED-LINE BEARDTONGUE, DEDALERA

White beardtongue grows on rocky, well-drained soil from Texas to the northern Great Plains. This perennial reaches two feet or more in height.

Although the seedpods take an exceptionally long time to mature, up to two months or more, white beardtongue reproduces easily from seed. Like its other *Penstemon* cousins, white beardtongue adapts readily to a home garden and is outstanding in a native landscape.

For more information about *Penstemon* characteristics and medicinal properties, see p. 220.

White beardtongue's field appearance and seedpods are shown on pp. 286 and 287.

SOLANACEAE (Nightshade Family)

Datura inoxia

DATURA, JIMSONWEED, SPINY APPLE

The large, showy flowers of the datura are at their best at night and in the early-morning hours, when their heavy narcotic perfume attracts night-flying hawk moths. The funnel-shaped white flowers are three to four inches wide and glow in the moonlight.

Datura makes an arresting garden plant; however, one should keep in mind that all parts of it are toxic. Even prolonged exposure to the scent can make a person feel nauseated or drowsy. The seeds, which are contained in a spiny capsule, are very hallucinogenic and cause severe sickness and death. Honey made from the nectar may be toxic.

The name "jimsonweed," or "Jamestown weed," is derived from a serendipitous military victory by American colonists against the British army in 1676. The Royalists ate a pot of wild datura greens that caused them to hallucinate for several days, allowing the Americans to take advantage of the British army's compromised condition.

Datura, Jimsonweed, Spiny Apple

Jimsonweed's highly poisonous seeds can be ground and mixed with tallow or kerosene as a method of exterminating body lice. The Pueblo people used the plant as an anodyne, or painkiller, for simple operations, including setting bones and repairing skin wounds. Smoke from burning leaves is said to be an excellent remedy for calming spasms during asthma attacks. The leaves can be used as a poultice for inflamed joints to relieve both pain and swelling, or they may be applied to the temples to relieve headache. Atropine, one of several highly dangerous alkaloids contained in jimsonweed, has medical applications and was extracted during World War I when there were shortages of that drug.

Some people have deliberately sought the hallucinogenic properties of datura. A tincture made from flowers of a South American species is said to induce clairvoyance. Several Native American tribes used datura as part of their manhood initiation rites, but it sometimes resulted in their untimely demise.

The spiny seedpod is shown on p. 287.

Daucus carota

QUEEN ANNE'S LACE

The delicate pattern of the tiny white flowers of Queen Anne's lace is reminiscent of the fine lace worn by the British monarch Queen Anne. In most plants there is a sterile black or dark purple flower in the center of the umbel, the cluster of diminutive flowers attached at a single point like an umbrella. Legend has it that it represents a drop of blood from Queen Anne's finger as she was sewing the lace. The leaves are delicate and dainty; in the days when hats adorned with feathers were the fashion, ladies used these leaves as a more affordable substitute.

A native of Europe, Queen Anne's lace now grows widely throughout the Western Hemisphere. Its ability to spread by seeds and roots has made it a noxious weed in some areas.

The seed head is almost as appealing as the flowers. As the seeds mature, the branchlets of the flower head curve up into a cuplike shape, referred to as a bird's nest. The seeds are covered with recurved barbs that cling to clothes or fur.

As the ancestor of the modern carrot, Queen Anne's lace has an edible root. The roots may be roasted to serve as a coffee substitute, and the seeds may be used to season soups. Rather surprisingly, the entire frilly cluster of flowers is good when battered and fried, but honestly, what isn't? However, because the plant resembles poison hemlock in appearance, do not ingest any parts of the plant if there is doubt about its identity! The juice can cause allergic reactions in some people, especially when the wet foliage is handled.

Early Spaniards in North America believed the root could serve as an antidote to rattlesnake bite. The seeds contain an oil that has a sedative effect and reduces spasms, acts as a hypoglycemic agent to lower blood

Daucus carota, *Queen Anne's lace; field view*   Daucus carota, *Queen Anne's lace; mature seeds*

sugar, and is an ingredient in face creams intended to decrease wrinkles. One novel use dates back to the fourth and fifth centuries BC, when Queen Anne's lace seeds were first used as a contraceptive; a progester-one-like compound in the oil is believed to prevent implantation of the embryo.

A field view of the plant, as well as the mature seed head, is shown here.

VISCACEAE (Mistletoe Family)

Phoradendron spp.

MISTLETOE

Mistletoe is a semiparasitic plant that inhabits the branches of several species of host trees, including oaks, mesquites, and other deciduous trees. It invades the stem tissue of the host with rootlike structures called haustoria, whose primary role is to obtain water, since mistletoe has chlorophyll for delivering its own food supply. Even so, mistletoe damage to the host plant may be drastic enough to cause the ends of branches to atrophy and die.

A number of toxic compounds are found in mistletoe, some of which have antitumor effects that are being further investigated. Extracts have also been shown to affect blood pressure and heart rate (so can a kiss under the mistletoe!).

Mistletoe is spread by birds, which eat the fruit without harm from the toxins, and disperse the seeds as they travel. It has also expanded its territory into West Texas and the Panhandle by piggy-backing on the

branches of young oak trees brought from Central and East Texas for
landscaping.

The history of mistletoe is rich in magical myths and beliefs. In
England, Druids kept mistletoe indoors as an abode for woodland spirits
in the cold of winter. They believed that its sacred nature was demonstrated by the way it grows in the air rather than in the earth, and harvesting it was thus accompanied by appropriate ceremonies. People were
given portions of the mistletoe to make rings and bracelets that they
believed would ward off evil spirits, witches, and disease.

The Saxons, who also inhabited the British Isles, named it *mistl-tan,*
which means "different twig," in reference to the fact that it is different from the plant on which it grows. An old Christian legend tells that
the mistletoe was once a tree, but when its wood was used to make the
cross, it shrank and became parasitic. Monks used to call it "wood of the
cross" and wore fragments around their necks to ward off disease.

A view of mistletoe parasitizing a tree is shown on p. 287.

From edible Yucca pods to tumbleweed seedlings, this section will enhance your appreciation of the range of personalities of white-flowering plants. Whole-plant views, some in their scenic context, also help in plant identification.

Dasylirion texanum, *sotol; leaves showing hooked spines on the leaf margins, p. 230.*

Dasylirion texanum, *sotol; male flowers of dioecious (meaning "two houses") sotol plants, with males and females borne on separate plants, p. 230.*

Dasylirion texanum, *sotol; field view, p. 230.*

Yucca glauca, *yucca; typical field appearance of plant with mature seedpods. During most of the year, dry flower stalks remain even as the plant flowers again the following season, p. 231.*

Yucca baccata, *banana yucca; field appearance with mature seedpods, p. 231.*

Yucca torreyi, *Torrey yucca; field appearance, p. 231.*

Yucca treculeana, *Spanish dagger; field appearance in chaparral desert habitat, p. 231.*

Asclepias engelmanniana, *narrow-leaved milkweed; dehisced seedpod releasing seeds with coma (tuft of fluff) attached, p. 233.*

Asclepias engelmanniana, *narrow-leaved milkweed; follicles (seedpods), p. 233.*

Rhus microphylla, *littleleaf sumac; note hairy fruits and pinnately compound leaves with winged rachis, p. 235.*

WHITE FLOWERS

Parthenium confertum, *false ragweed; field appearance, p. 242.*

Salsola tragus, *tumbleweed; field appearance of mature plant—too tough to chew, but ready to roll! p. 245.*

Merremia dissecta, *Alamo vine; young seedpods, p. 249.*

Opuntia leptocaulis, *Christmas cactus; red fruits decorate stems most of the year, p. 243.*

Salsola tragus, *tumbleweed; plant is edible at this seedling stage, p. 245.*

Convolvulus equitans, *bindweed; seedpods showing remnant of style on the mature ovary—a Hershey kiss! p. 247.*

Merremia dissecta, *Alamo vine; deeply lobed leaf with palmate venation, p. 249.*

Cnidoscolus texanus, *bull nettle; all plant parts are armed with stinging hairs, p. 253.*

Euphorbia bicolor, *snow-on-the-prairie; note the strongly exserted pendulous green ovaries, typical of the Euphorbiaceae family, p. 254.*

Astragalus racemosus, *alkali milkvetch; note angular bean pods and pinnately compound leaves, p. 255.*

Dalea candida, *white prairie clover; field appearance, p. 256.*

Desmanthus illinoensis, *Illinois bundle flower; note clusters of dehiscing mature bean pods and bipinnately compound leaves, p. 257.*

Allium drummondii, *wild onion; flower parts in multiples of three, p. 262.*

Allium drummondii, *wild onion; mature fruit with seeds, p. 262.*

Mentzelia decapetala, *tenpetal blazingstar;*
two mature fruits and one bud, p. 264.

Eriogonum alatum *var.* glabriusculum,
grassland buckwheat; field appearance,
p. 271.

Eriogonum alatum *var.* glabriusculum,
grassland buckwheat; close-up of flowers with
winged pendulous fruit, p. 271.

Sapindus saponaria *var.* drummondii,
soapberry; fruit showing the translucency of
the berries, p. 276.

Penstemon albidus, *white penstemon; field*
appearance, p. 277.

Datura inoxia, *jimsonweed; young seedpod with spiny protuberances, p. 277.*

Phoradendron spp., *mistletoe; appearance of mistletoe ensconced in its host tree, p. 280.*

Penstemon albidus, *white penstemon; young seedpods with filamentous styles still attached, p. 277.*

Ajilvsgi, Geyata. 2003. *Wildflowers of Texas*. Rev. ed. Fredericksburg, Tex.: Shearer Publishing.

Alarcon-Aguilar, F. J., F. Calzada-Bermejo, E. Hernandez-Galicia, C. Ruiz-Angeles, and R. Roman-Ramos. 2005. "Acute and Chronic Hypoglycemic Effect of *Ibervillea sonorae* Root Extracts—II." *Journal of Ethnopharmacology* 97:447–52. Included in Diabetes Medicinal Plant Database of ProGene Biosciences. http://www.progenebio.in/DMP/listz.htm (site updated March 2006; accessed July 15, 2008).

Ancient/Classical History Glossary. 2008. http://ancienthistory.about.com/od/classics/a/glossary.htm (accessed July 15, 2008).

Armstrong, Wayne. *An Online Textbook of Natural History*. San Marcos, Calif.: Life Science Department, Palomar College. http://waynesword.palomar.edu/indxwayn.htm (site updated May 1, 2008; accessed July 15, 2008).

Austin, D. F. 1997. "Convolvulaceae (Morning Glory Family)." http://www.fau.edu/biology/people/daustin/convolv.htm (site updated October 2001; accessed July 15, 2008).

Back, Philippa. 1994. *The Illustrated Herbal*. London: Reed International Books.

Beasley, V., ed. 1999. *Toxicants That Inhibit the Function of Cytochromes. Veterinary Toxicology*. Ithaca, N.Y.: International Veterinary Information Service. http://www.ivis.org/advances/Beasley/Cpt17/IVIS.pdf (accessed July 15, 2008).

Benson, Lyman. 1959. *Plant Classification*. Lexington, Mass.: D. C. Heath.

Bremness, Lesley, contributing ed. 1990. *Reader's Digest Home Handbooks, Herbs*. Pleasantville, N.Y.: Reader's Digest Association.

Brown, Lauren. 1979. *Grasses: An Identification Guide*. New York: Houghton Mifflin.

Brunelle, Paul. *Economics of Cacti*. 1999. Dalhousie Collection of Cacti & Other Succulents, Dalhousie University, Nova Scotia. http://cactus.biology.dal.ca/economics.html (site updated August 24, 2004; accessed July 15, 2008).

Bunney, Sarah. 1984. *The Illustrated Encyclopedia of Herbs*. New York: Dorset Press.

Charters, Michael L. 2003–2004. "California Plant Names: Latin Name Meanings and Derivations." http://www.calflora.net/botanicalnames/ (site updated July 1, 2008; accessed July 15, 2008).

Cheatham, Scooter, Marshall C. Johnston, and Lynn Marshall. 1995. *Useful Wild Plants of Texas, the Southeastern and Southwestern United States, the Southern Plains, and Northern Mexico*. Vol. 1, *Abronia* through *Arundo*. Austin: Useful Wild Plants.

Correll, Donovan Stewart, and Marshall Conring Johnston. 1970. *Manual of the Vascular Plants of Texas*. Renner: Texas Research Foundation.

Cox, Jeff, and Marilyn Cox. 1985. *The Perennial Garden: Color Harmonies through the Seasons*. Emmaus, Penn.: Rodale Press.

Crawford, Hester M., Frederick McGourty, and Marjorie J. Dietz, eds. 1987. *Herbs and Their Ornamental Uses. Plants and Gardens, Brooklyn Botanic Garden Record*. Brooklyn, N.Y.: Brooklyn Botanic Garden.

Damron, B. L., and J. P. Jacob. 2001. *Toxicity to Poultry of Common Weed Seeds*. Fact Sheet PS-55. Animal Sciences Department, Florida Cooperative Extension Service, Institute of Food and Agricultural Sciences, University of Florida. http://edis.ifas.ufl.edu/PS052 (accessed July 15, 2008).

Deng, Yu-Cheng, Hui-Ming Hua, Jun Li, and Peter Lapinskas. 2001. "Studies on the Cultivation and Uses of Evening Primrose (*Oenothera* spp.) in China." *Economic Botany* 55 (1): 83–92. http://www.lapinskas.com/publications/3586.html (site updated July 2003; accessed July 15, 2008).

Dieringer, Gregg, and Leticia Cabreara. 1994. "Sexual Selection of Anther Trichomes and Sexual Dimorphism in *Ibervillea lindheimeri* (Cucurbitaceae: Melothrieae)." *American Journal of Botany* 81 (1): 111–18.

Duke, James. 1994. *Phytochemical and Ethnobotanical Databases*. Agricultural Research Services, USDA. http://www.ars-grin.gov/duke/plants.html (site updated March 10, 1998; accessed July 15, 2008).

———. 1997. *The Green Pharmacy*. New York: St. Martin's Press.

Dunmire, W. W., and Gail D. Tierney. 1997. *Wild Plants and Native Peoples of the Four Corners*. Santa Fe: Museum of New Mexico Press.

Elias, Thomas S., and Peter A. Dykeman. 1990. *Edible Wild Plants: A North American Field Guide*. New York: Sterling Publishing.

Emerson, Julia T., and William H. Welker. 1908. "Some Notes on the Chemical Composition and Toxicity of *Ibervillea sonorae*." *Journal of Biological Chemistry* 5:339–50. http://www.jbc.org/cgi/reprint/5/4/339.pdf (accessed July 15, 2008).

Enquist, Marshall. 1987. *Wildflowers of the Texas Hill Country*. Austin: Lone Star Botanical.

Everitt, James H., and D. Lynn Drawe. 1993. *Trees, Shrubs & Cacti of South Texas*. Lubbock: Texas Tech University Press.

Faucon, Philippe. 2003. http://www.desert-tropicals.com/Plants/sci_names.html (site updated April 29, 2007; accessed July 15, 2008).

Ferrigni, N. R., D. E. Nichols, J. L. McLaughlin, and R. A. Bye. 1982. "Cactus Alkaloids. XLVII. N alpha-dimethylhistamine, a Hypotensive Component of *Echinocereus triglochidiatus*." *Journal of Ethnopharmacology* 5 (3): 359–64. http://www.ncbi.nlm.nih.gov/entrez/query.fcgi?cmd=Retrie

ve&db=PubMed&list_uids=7087506&dopt=Abstract (site accessed July 15, 2008).

Fetrow, Charles W., and Juan R. Avila. 1999. *Professional's Handbook of Complementary and Alternative Medicines*. Springhouse, Penn.: Springhouse.

Fire Effects Information System. U.S. Department of Agriculture, Forest Service, Rocky Mountain Research Station, Fire Sciences Laboratory (producer). http://www.fs.fed.us/database/feis (site updated July 11, 2008; accessed July 15, 2008).

Forey, Pamela. 1998. *Wild Flowers of North America*. North Dighton, Mass.: JG Press.

Francis, John K. "*Tecoma stans*. (L.) Juss. ex HBK." Río Piedras, Puerto Rico: International Institute of Tropical Forestry, U.S. Department of Agriculture, Forest Service. http://www.fs.fed.us/global/iitf/Tecoma%20stans.pdf (accessed July 15, 2008).

Gardening Plant Information. 2004. GreenWeb Company. http://www.boldweb.com/gw (site updated August 2004; accessed July 15, 2008).

Giannangelo, Frank, and Vicky Giannangelo. 2008. Giannangelo Farms Southwest. http://www.avant-gardening.com/New%20Mexico.html (accessed July 15, 2008).

Gibson, Arthur C. Economic Botany: Writeups and Illustrations of Economically Important Plants. University of California at Los Angeles. http://www.botgard.ucla.edu/html/botanytextbooks/economicbotany/index.html (accessed July 15, 2008).

Global Herbal Supplies. 2008. The Herbal Information Library. http://www.globalherbalsupplies.com/herb_information/ (accessed July 15, 2008).

Gould, Frank. 1975. *The Grasses of Texas*. College Station: Texas A&M University Press.

Grieve, Maud. 1996. *A Modern Herbal*. New York: Barnes and Noble Books (orig. pub. 1931). www.botanical.com/botanical/mgmh/mgmh.html (accessed July 15, 2008).

A Guide to Medicinal and Aromatic Plants. 2000. West Lafayette, Ind.: Center for New Crops and Plant Products, Department of Horticulture and Landscape Architecture, Purdue University. http://www.hort.purdue.edu/newcrop/Indices/index_ab.html (site updated December 18, 2007; accessed July 15, 2008).

Haddock, Mike. 2008. Kansas Wildflowers and Grasses. Kansas State University. http://www.lib.ksu.edu/wildflower/ (site updated June 23, 2008; accessed July 15, 2008).

Harput, U. S., Iclal Saracoglu, and Yukio Ogihara. 2004. "Methoxyflavonoids from *Pinaropappus roseus*." *Turkish Journal of*

Chemistry 28:761–66. http://journals.tubitak.gov.tr/chem/issues/kim-04-28-6/kim-28-6-11-0405-2.pdf (accessed July 15, 2008).

Harris, Ben Charles. 1961. *Kitchen Medicines*. Worcester, Mass.: Nature Publications.

Hart, Charles, Tam Garland, Catherine Barr, Bruce Carpenter, and John Reagor. 2004. *Toxic Plants of Texas*. College Station: Texas Cooperative Extension.

Hatch, Stephan L., and Jennifer Pluhar. 1993. *Texas Range Plants*. College Station: Texas A&M University Press.

Haukos, David A., and Loren M. Smith. 1997. *Common Flora of the Playa Lakes*. Lubbock: Texas Tech University Press.

Health Claims. 1994. A Food Labeling Guide—Appendix C. U.S. Food and Drug Administration. Center for Food Safety and Applied Nutrition. http://www.cfsan.fda.gov/~dms/flg-6c.html#upd (site updated April 11, 2008; accessed July 15, 2008).

Hemphill, Rosemary. 1972. *Herbs for All Seasons*. Harmondsworth, Middlesex, U.K.: Penguin Books.

Herbs2000. 2008. http://www.herbs2000.com/herbs/1menu.htm (accessed July 15, 2008).

Historical Development of the Ashmolean. 2005. University of Oxford, Ashmolean Museum. http://www.ashmolean.org/about/historyandfuture/ (site updated 2005; accessed July 15, 2008).

Hitchcock, A. S. 1950. *Manual of the Grasses of the United States*. 2nd ed., rev. by Agnes Chase. Misc. Pub. No. 200. Washington, D.C.: United States Department of Agriculture, U.S. Government Printing Office.

Index of Herbal Medicines and Supplements. 2008. Aetna InteliHealth. http://www.intelihealth.com/IH/ihtIH/WSIHW000/8513/8513.html (site updated August 29, 2005; accessed July 15, 2008).

Industrial Uses of Agricultural Materials. 1996. Economic Research Service. USDA. http://www.bioplastic.org/industrial-use-1996.html (accessed July 15, 2008).

International Institute of Tropical Forestry. USDA Forest Service. http://www.fs.fed.us/global/iitf/pdf/shrubs/ (site updated March 29, 2004; accessed July 15, 2008).

James, Lynn, Kay Lynn Bennett, Karl G. Parker, Richard F. Keeler, Wayne Binns, and Ben Lindsay. 1968. "Loco Plant Poisoning in Sheep." *Journal of Range Management* 21 (6): 360–65.

Kane, Charles. 2006. *Herbal Medicine of the American Southwest: A Guide to the Identification, Collection, Preparation, and Use of Medicinal and Edible Plants of the Southwestern United States*. Tucson, Ariz.: Lincoln Town Press.

Kantrud, Harold A. 1995. *Native Wildflowers of the North Dakota Grasslands*. Jamestown, N.D.: Northern Prairie Wildlife Research Center Online.

http://www.npwrc.usgs.gov/resource/plants/wildflwr/index.htm
(Version 06JUL2000) (site updated August 24, 2006; accessed July 15, 2008).

Kindscher, Kelly. 1987. *Edible Wild Plants of the Prairie*. Lawrence: University Press of Kansas.

————. 1992. *Medicinal Wild Plants of the Prairie*. Lawrence: University Press of Kansas.

Kirkpatrick, Zoe Merriman. 1992. *Wildflowers of the Western Plains*. Austin: University of Texas Press.

Knott, Ron. 1996–2003. "Fibonacci Numbers and Nature." http://www.mcs .surrey.ac.uk/Personal/R.Knott/Fibonacci/fibnat.html#petals (site updated March 28, 2008; accessed July 15, 2008).

Koerper, H. C., and N. A. Whitney-Desautels. 1999. "*Astragalus* Bones: Artifacts or Ecofacts?" *Pacific Coast Archaeological Society Quarterly* 35 (2–3): 69–80. http://www.pcas.org/Vol35N23/3523Koerper.pdf (accessed July 15, 2008).

Kowalchik, Claire, and William H. Hylton, eds. 1987. *Rodale's Illustrated Encyclopedia of Herbs*. Emmaus, Penn.: Rodale Press.

Lacy, Allen. 1986. *Farther Afield: A Gardener's Excursions*. New York: Farrar, Straus Giroux.

Leonard, David B. 2008. Medicine at Your Feet. http://www .medicineatyourfeet.com/plantindex.html (accessed July 15, 2008).

Levy, Juliette de Bairacli. 1982. *The Illustrated Herbal Handbook*. London: Thetford Press.

Lima, Patrick. 1988. *The Harrowsmith Northern Perennials Handbook*. Charlotte, Vt.: Camden House Publishing.

Lloyd, J. U. 1898. Croton tiglium. Reprinted from *The Western Druggist*, Chicago. http://www.swsbm.com/ManualsOther/Croton%20tiglium-Felter.pdf (accessed July 15, 2008).

Loughmiller, Campbell, and Lynn Loughmiller. 1984. *Texas Wildflowers*. Austin: University of Texas Press.

Lucas, Richard. 1966. *Nature's Medicine*. West Nyack, N.Y.: Parker Publishing.

Marichal, Carlos. 2001. "A Forgotten Chapter of International Trade: Mexican Cochineal and the European Demand for American Dyes, 1550–1850." Paper presented to the Conference on Latin America Global Trade and International Commodity Chains in Historical Perspective, Stanford University, November 16–17. http://sshi.stanford .edu/Conferences/2001-2002/GlobalTrade2001/marichal.pdf (accessed July 15, 2008).

Mattiza, Dorothy Baird. 1993. *100 Texas Wildflowers*. Tucson, Ariz.: Southwest Parks and Monuments Association.

McDonald, R. W., W. Bunjobpan, T. Liu, S. Fessler, O. E. Pardo, Ik. Freer, M. Glaser, M. J. Seckl, and D. J. Robins. 2001. "Synthesis and Anticancer Activity of Nordihydroguaiaretic Acid (NDGA) and Analogues." *Anticancer Drug Design* 16 (6): 261–70. http://www.ncbi.nlm.nih.gov/ entrez/query.fcgi?cmd=Retrieve&db=PubMed&list_uids=12375879&do pt=Abstract (accessed July 15, 2008).

McGourty, Frederick, ed. 1987. *Herbs and Their Ornamental Uses.* New York: Brooklyn Botanic Garden.

McLain-Romero, J., R. Creamer, H. Zepeda, J. Strickland, and G. Bell. 2004. "The Toxicosis of *Embellisia* Fungi from Locoweed (*Oxytropis lambertii*) Is Similar to Locoweed Toxicosis in Rats." *Journal of Animal Science* 82:2169–74. http://jas.fass.org/cgi/reprint/82/7/2169.pdf (accessed July 15, 2008).

McMahan, Craig, and Jack M. Inglis. 1974. "Use of Rio Grande Plain Brush Types by White-tailed Deer." *Journal of Range Management* 27 (5): 369–74.

Medicinal Plants of Manitoba. Manitoba Agriculture, Food and Rural Initiatives. http://www.gov.mb.ca/agriculture/crops/medicinal/ bkq00s00.html (site updated March 2006; accessed July 15, 2008).

Mifsud, Stephen. 2007. "Wild Plants of the Mediterranean Islands of Malta." www.MaltaWildPlants.com (site updated March 10, 2008; accessed July 15, 2008).

Mitich, Larry W. 2001. "Intriguing World of Weeds." Weed Science Society of America. http://www.wssa.net/Weeds/ID/WorldofWeeds.htm (site updated 2008; accessed March 15, 2008).

Moore, Michael. 1979. *Medicinal Plants of the Mountain West.* Santa Fe: Museum of New Mexico Press.

———. 1989. *Medicinal Plants of the Desert and Canyon West.* Santa Fe: Museum of New Mexico Press.

———. 1990. *Los Remedios: Traditional Herbal Remedies of the Southwest.* Santa Fe: Red Crane Books.

Native American Ethnobotany. A Database of Foods, Drugs, Dyes and Fibers of Native American Peoples, Derived from Plants. University of Michigan, Dearborn. http://herb.umd.umich.edu/ (accessed July 15, 2008).

Native Plant Database, Lady Bird Johnson Wildflower Center. University of Texas at Austin. http://www.wildflower.org/plants/ (accessed July 15, 2008).

Niering, William A., and Nancy C. Olmstead. 1979. *The Audubon Society Field Guide to North American Wildflowers, Eastern Region.* New York: Alfred A. Knopf.

Niethammer, Carolyn. 1974. *American Indian Food and Lore: 150 Authentic Recipes*. New York: Collier Books, Macmillan Publishing.

———. 1987. *The Tumbleweed Gourmet: Cooking with Wild Southwestern Plants*. Tucson: University of Arizona Press.

Northern Prairie Wildlife Research Center. U.S. Department of the Interior, U.S. Geological Survey, Jamestown, N.D. http://www.npwrc.usgs.gov/ (site updated April 16, 2007; accessed July 15, 2008).

O'Connell, Mary. "Medicinal Plants of the Southwest." Department of Agronomy & Horticulture, New Mexico State University. http://medplant.nmsu.edu (site updated February 13, 2008; accessed July 15, 2008).

Opinion of the Scientific Committee on Food on the 19th Additional List of Monomers and Additives for Food Contact Materials. 2002. European Commission Health & Consumer Protection Directorate–General Scientific Committee on Food. Brussels, Belgium. http://europa.eu.int/comm/food/fs/sc/scf/out141_en.pdf (accessed July 15, 2008).

Parker, Kittie F. 1972. *An Illustrated Guide to Arizona Weeds*. Tucson: University of Arizona Press. http://www.uapress.arizona.edu/onlinebks/weeds/titlweed.htm (accessed July 15, 2008).

Phillips, Ellen, and C. Colston Burrell. 1993. *Illustrated Encyclopedia of Perennials*. Emmaus, Penn.: Rodale Press.

Plants for a Future, Species Database. 1997–2001. Lostwithiel, Cornwall, UK: The Field, Penpol. http://www.ibiblio.org/pfaf/D_search.html (site updated February 2002; accessed July 15, 2008).

Plants of Wisconsin. Robert W. Freckmann Herbarium, University of Wisconsin, Stevens Point, Wisconsin. http://wisplants.uwsp.edu/WisPlants.html (accessed July 15, 2008).

Plants Poisonous to Livestock in the Western States. Agriculture Information Bulletin 415. Poisonous Plant Research Laboratory, Agricultural Research Services, Logan, Utah: USDA. http://www.pprl.ars.usda.gov/ (site updated February 8, 2006; accessed July 15, 2008).

Poisonous Plants. Faculty of Pharmacy, Assiut University, Egypt. http://www.aun.edu.eg/distance/pharmacy/poison/poisonous_plants.htm (accessed July 15, 2008).

Poison Plants Information Database. Department of Animal Science, Cornell University. http://www.ansci.cornell.edu/plants/index.html (site updated January 11, 2008; accessed July 15, 2008).

Powell, A. Michael. 1994. *Grasses of the Trans-Pecos and Adjacent Areas*. Austin: University of Texas Press.

Ranson, Nancy Richey. 1933. *Texas Wild Flower Legends*. Dallas: Kaleidograph Press.

Richardson, Alfred. 1995. *Plants of the Rio Grande Delta*. Austin: University of Texas Press.

Romo, J. T. 2000. *Rangeland Ecosystems and Plants*. Fact Sheets for Some Common Plants on Rangelands in Western Canada. Department of Plant Science, University of Saskatchewan. http://www.usask.ca/agriculture/plantsci/classes/range/index.html (accessed July 15, 2008).

Rose, Francis L., and Russell W. Strandtmann. 1986. *Wildflowers of the Llano Estacado*. Dallas: Taylor Publishing.

Sanders, Jack. 1995. *Hedgemaids and Fairy Candles*. Camden, Maine: Ragged Mountain Press.

Savinelli, Alfred. 1997. *Plants of Power*. Taos, N.M.: Alfred Savinelli.

Schulz, Ellen D. 1928. *Texas Wild Flowers*. Chicago: Laidlaw Brothers.

Silverthorne, Elizabeth. 1996. *Legends and Lore of Texas Wildflowers*. College Station: Texas A&M University Press.

Simpson, Benny J. 1988. *A Field Guide to Texas Trees*. Austin: Texas Monthly Press.

Spellenberg, Richard. 1979. *The Audubon Society Field Guide to North American Wildflowers*. New York: Alfred A. Knopf.

Summary of All GRAS Notices. 2004. Center for Food Safety and Applied Nutrition, Office of Food Additive Safety, U.S. Food and Drug Administration. http://vm.cfsan.fda.gov/~rdb/opa-gras.html (site updated June 26, 2008; accessed July 15, 2008).

Swerdlow, Joel L. 2000. *Nature's Medicine, Plants That Heal*. Washington, D.C.: National Geographic Society.

Texas Native Shrubs. Texas Ornamentals Research and Education Program, Texas A&M University. http://aggie-horticulture.tamu.edu/ornamentals/nativeshrubs/indexscientific.htm (accessed July 15, 2008).

Thomson Healthcare. PDR Health Drug Information. 2008. http://www.pdrhealth.com/drugs/altmed/altmed-a-z.aspx (accessed July 15, 2008).

Toxicological Profile for Wood Creosote, Coal Tar Creosote, Coal Tar, Coal Tar Pitch, and Coal Tar Pitch Volatiles. 2002. Atlanta: U.S. Department of Health and Human Services, Public Health Service, Agency for Toxic Substances and Disease Registry. www.atsdr.cdc.gov/toxprofiles/tp85.pdf (accessed July 15, 2008).

Tull, Delena. 1987. *A Practical Guide to Edible and Useful Plants*. Austin: Texas Monthly Press.

Turner, Billie L., H. Nichols, G. Denny, and O. Doron. 2003. *Atlas of the Vascular Plants of Texas*. Fort Worth: Botanical Research Institute of Texas.

Turner, Sharon. 2001. *Witwatersrand National Botanical Garden*. South African National Biodiversity Institute. http://www.plantzafrica.com (accessed July 15, 2008).

United States Congress, Office of Technology Assessment. Selected
New Industrial Crops, Appendix A. 1991. Agricultural Commodities
as Industrial Raw Materials. http://www.princeton.edu/~ota/
disk1/1991/9105/9105.PDF (accessed July 15, 2008).

Vogel, Virgil. 1970. *American Indian Medicine*. Norman: University of
Oklahoma Press.

Warnock, Barton H. 1970. *Wildflowers of the Big Bend Country, Texas*. Alpine,
Tex.: Sul Ross State University.

Wasowski, Sally, and Andy Wasowski. 1991. *Native Texas Plants:
Landscaping Region by Region*. Houston: Gulf Publishing.

Welch, William C. 2002. "Texas Mountain Laurel, *Sophora secundiflora*." In
Horticulture Update, April 2002. College Station: Extension Horticulture,
Texas Cooperative Extension, Texas A&M University System. http://
aggie-horticulture.tamu.edu/extension/newsletters/hortupdate/apr02/
art3apr.html (accessed July 15, 2008).

Wells, Diana. 1997. *100 Flowers and How They Got Their Names*. Chapel Hill,
N.C.: Algonquin Books of Chapel Hill.

Whitson, Tom D., Larry C. Burrill, Steven A. Dewey, David W. Cudney,
B. E. Nelson, Richard D. Lee, and Robert Parker, eds. 2008. *Weeds of the
West*. 10th ed. Western Society of Weed Science in cooperation with the
Western United States Land Grant Universities Cooperative Extension
Service. Las Cruces, N.M.: Western Society of Weed Science.

Yarris, Lynn. 2004. "Synthetic Biology Offers New Hope for Malaria
Victims." Science Beat, Berkeley Lab. http://www.lbl.gov/Science-
Articles/Archive/sb-PBD-anti-malarial.html (accessed July 15, 2008).

INDEX

Page numbers in **bold** refer to each plant's main entry.

A'Court, Mary Elizabeth, 187
Abronia angustifolia, **208–209**, 226
achene, 16, 66, 113, 181
Achillea millefolium, **237–238**
Acourtia nana, **187**
agarita, 2, **76–77**, 114
Agavaceae (family characteristics), 2
Agave, 2
Agave Family (family characteristics), 2
ageratum, wild, **150–151**
agrito (*Berberis trifoliolata*), 2, **76–77**, 114
agrito (*Rhus microphylla*), 2, **235–236**, 283
Alamo vine, **249–250**, 284
alcoholic beverage, 88, 231, 261
alegria, 2, **120–121**
alfalfa, **159–160**
algerita, 2, **76–77**, 114
alkali milkvetch, **255**, 285
alkaline soil, 5, 31, 126, 176, 241
alkaloid, 9, 25, 44, 53, 67, 74, 77, 91, 98, 105, 107, 129, 138, 140, 142, 157, 160, 161, 162, 163, 175, 219, 255, 273, 278
allelopathy, 122
allergy (allergic reaction, allergen), 57, 74, 104, 159, 279
Allioni, Carlos, 209
Allionia incarnata, **209–210**
allionia, trailing, **209–210**
Allium drummondii, **262–263**, 285
Amaranthaceae (family characteristics), 3
Amaranthus spp., 2, **120–121**
Amblyolepis setigera, **48**
Ambrosia grayi, **121–122**, 125
anabasine, 105
Anacardiaceae (family characteristics), 3
Anemone berlandieri, **272–273**

anemone, tenpetal, **272–273**
angel hair, **247–249**
añil del muerto, **74**, 113
annual lupine, **142**
annual widow's tears, **139**
annual wild bean, **199–200**, 225
anodyne, 278
antelope horns, 2, **233–235**
anther, 9, 30, 63, 65, 79, 115, 139, 153, 165, 209, 211, 214, 221, 233, 249
anthocarps, 34, 169, 182
anthraquinone, 86
antibacterial properties, 44, 77, 123, 138, 170, 233, 258
antifungal, 23, 123, 193, 233, 258
antihelminthic agent, 123
anti-inflammatory agent, 60, 74, 77, 102, 104, 191, 237, 261
antimicrobial properties, 74, 273
antioxidants, 104, 243
antiseptic, 60, 100, 129, 204, 243
antitumor agent, 57, 74, 104, 109, 280
antiviral agent, 78, 168, 191
Apache Indians, 25, 56, 72, 77, 79, 102, 133, 151, 197, 268, 275
Apache tea, 5, **78–79**
Apache-plume, 17, **275**
Aphanostephus skirrhobasis, **238**
Aquilegia hinckleyana, 17, **101**, 116
Argemone mexicana, **98–99**
Argemone polyanthemos, **268–269**
Argemone sanguinea, **213–214**
arid land adaptations (drought tolerance), 4, 77, 22, 27, 28, 50, 63, 64, 85, 151, 154, 168, 179, 187, 205, 208, 243, 258, 261, 265
Arizona skipper butterfly, 258
Arkansas lazy daisy, **238**
Artemisia filifolia, **123–124**, 133
Artemisia ludoviciana, **123–124**, 133
arthritis, 56, 60, 81, 94, 130, 160, 164, 217, 232, 265, 268, 276
asadero, 175
asclepain, 234

Blackfoot (Blackfeet) Indians, 23, 54,
56, 151, 234, 262
blackfoot daisy, **241–242**
bladder infection, 27, 43, 52, 99, 104,
250, 258, 266
bladderpod, 6, **82**
blazing star, **151–152**, 181
blazingstar, tenpetal, 13, **264–265**,
286
Bleeding Heart Family (family
characteristics), 9
blind staggers, 157
blood pressure, 92, 110, 138, 170, 250,
280
bloodstanch, **124**
blue boneset, **150–151**
blue curls, **165**
blue curls, gyp, 11, **164–165**
blue flax, 11, **145–146**
blue gilia, 15, **146**, 147
blue sage, **145**
bluebell, 9, **143**
bluebonnet, 9, **141–142**, 147
bluebonnet, Big Bend, **142**, 147
bluebonnet, Texas, 9, **141–142**, 147
blue-eyed grass, 11, **140–141**, 147
blueweed, **58–59**, 112
boneset, blue, **150–151**
boneset, false, **240**
boneset, late-flowering, **240**
Boraginaceae (family characteristics), 4
bract milkweed, **150**, 181
bracts, 3, 10, 33, 54, 111, 116, 150, 176,
178, 187, 188, 189, 241, 254
breakbone fever (dengue fever), 150
bronchitis, 55, 73, 234
broomweed, **55-56**, 111
Brown-eyed Susan, 2, **65–66**
bruisewort, **239**
buckeye, Mexican, **218–219**, 226
Buckthorn Family (family
characteristics), 16
buckwheat, **271–272**
Buckwheat Family (family
characteristics), 15–16

buckwheat, grassland, **271–272**, 286
buckwheat, heartsepal, **271–272**
buffalo bur, **107–108**
buffalo clover, 9, **141–142**, 147
buffalo gourd, 6, 7, **83–84**, 114
bull nettle, **253**, 285
bunch-grass, **267–268**
bur ragweed, **121–122**, 125
bursage, woollyleaf, **121–122**, 125
bush morning-glory, **196**
bush-clover, slender, **197–198**
Buttercup Family (family
characteristics), 16
butterflies, 29, 38, 48, 75, 76, 86, 101,
140, 152, 154, 158, 165, 166, 172,
189, 192, 198, 200, 208, 211, 220,
221, 235, 258, 260, 265
butterfly daisy, 48
butterfly, Arizona skipper, 258
butterfly, dotted checkerspot, 172
butterfly, gray hairstreak, 265
butterfly-weed, **38**

Cactaceae (family characteristics), 4
Cactus Family (family characteristics),
4
cactus, Christmas, **243–244**, 284
cactus, claret cup, **25**, 34
cactus, devil's head, **193**, 225
cactus, dumpling, **24-25**
cactus, hedgehog, **25**, 34
cactus, horse crippler, **193**, 225
cactus, jumping, **243–244**, 284
cactus, prickly pear, 5, **80–81**, 114
cactus, red plains prickly pear, **26**
cactus, Runyon's nipple, **24-25**
cactus, southwestern barrel, **79–80**
cadillos, **125–126**
Caesalpinioideae, 9
calabazilla, 6, 7, **83–84**, 114
caliche, 42, 75, 146, 171, 256
caliche globemallow, 13, **42-43**
Callirhoe involucrata var. involucrata,
205–206, 226

marigold, fetid, **51**

matchweed, **55-56**, 111

Maximilian sunflower, **59–60**

Mayapple, **28–29**

maypop, 15, **169–170**

meadow flax, 11, **145–146**

mealy sage, **144**

Medicago sativa, **159–160**

Melampodium leucanthum, **241–242**

Melilotus albus, **88**

Melilotus officinalis, **88–89**

melon, coyote, **84–85**, 115

menopause, 92, 201

menstruation, 4, 13, 28, 60, 97, 154, 159, 164, 167, 179, 237, 250, 272

Mentzelia decapetala, 13, **264–265**, 286

Mentzelia nuda, **264–265**

Mentzelia reverchonii, **94**

mentzelia, tenpetal, 13, **264–265**, 286

Merremia dissecta, **249–250**, 284

mesquite, 20, **89–90**, 115, 235, 255, 280

meteorism (flatulence), 124, 262

methyl salicylate, 271

Mexican apple, **28–29**

Mexican buckeye, **218–219**, 226

Mexican gold poppy, **43-44**, 45

Mexican hat, **39**, 45

Mexican poppy, **98–99**

migraines, 99, 105

milfoil, **237–238**

milkvetch, alkali, **255**, 285

milkvetch, cream, **255**, 285

milkvetch, raceme, **255**, 285

Milkweed Family (family characteristics), 3

milkweed, bract, **150**, 181

milkweed, common, **186**, 224

milkweed, lone star, **150**, 181

milkweed, narrow-leaved, **234-235**, 283

milkweed, orange, **38**

milkweed, purple-flowered, **150**, 181

milkweed, showy, **186**, 224

Milkwort Family (family characteristics), 15

milkwort, white, 15, **270–271**

milky sap, 3, 8, 9, 14, 15, 38, 68, 69, 137, 153, 233, 235, 255

mimbre, 5, **192–193**, 225

Mimosa nuttallii, **199**

mimosa, prairie, **257**, 285

Mimosoideae, 9

Mint Family (family characteristics), 11

mint, lemon, **166–167**

Mirabilis glabrifolia, **168–169**, 182, 183

Missouri violet, 19, **180**

mistflower, **150–151**

mistletoe, 20, **280–281**, 287

Mistletoe Family (family characteristics), 20

Monarda citriodora, **166–167**

Monarda clinopodioides, **204**

Monarda pectinata var. punctata, **260–261**

Monarda punctata, **260–261**

Monardes, Nicholas, 166, 260

monk's pepper, **179**

monocot, 6, 10, 18, 139, 147, 249, 263

monoecious, 7, 20, 83, 122

monotypic genus, 93

mordant, 22, 66

Mormon tea, 6, **128-129**, 134

Morning-Glory Family (family characteristics), 7

morning-glory, big-root, **196**

morning-glory, bush, **196**

morning-glory, heavenly blue, 6, **139–140**

morning-glory, wild, **195**

mosquitoes, 162, 221, 251

moth mullein, **221–222**

moths, 32, 172, 203, 209, 213, 220, 221, 231, 232, 277

mountain pink, 8, **201–202**

mucilage (mucilaginous), 6, 12, 80, 94, 100, 138, 155, 205, 207, 266

mugwort, **123–124**, 133

rash, poison ivy, 39, 55, 175, 200, 252, 269

Ratany Family (family characteristics), 10

ratany, range, 11

ratany, trailing, **203**, 225

Ratibida columnaris, **39**, 45

Ratibida columnifera, **39**, 45

rattleweed, **156–158**, 181

ray flowers, 4, 23, 39, 56, 70, 72, 112, 189, 239, 241

rayed palafoxia, **191–192**, 225

rayless goldenrod, **61**

rayless greenthread, **72**

receptacle, 3

red clover, **201**

red plains prickly pear cactus, **26**

red poppy (*Argemone sanguinea*), **213–214**

red poppy (*Papaver rhoeas*), **30–31**, 34

red-line beardtongue, **277**, 286

red-neck evening primrose, **96–97**, 115, 116

resin (gum), 43, 89

respiratory problems, 8, 56, 91, 104, 106, 201, 274

Reverchon's stickleaf, **94**

Rhamnaceae (family characteristics), 16

rheumatism, 38, 58, 83, 101, 128, 160, 166, 211, 239, 268, 272

Rhizobium trifolii, 259

rhizome, 10, 18, 20, 58, 59, 75, 91, 122, 123, 133, 154, 174, 188, 189, 191, 195, 213, 240

Rhus, 3

Rhus microphylla, 2, **235–236**, 283

ringworm, 69, 109, 193

Rio Grande Valley Indians, 83

rituals, 25, 140, 163

river hemp, **40-41**, 45

rock centaury, 8, **201–202**

rock rose, **208**, 226

rock-lettuce, **243**

Rocky Mountain bee plant, 5, **244–245**

Roemer, Ferdinand, 90

Rosaceae (family characteristics), 16

Rose Family (family characteristics), 16

rose prickly poppy, **213–214**

rosette, 10, 141, 163

rosinweed, **54-55**

rosita, **201–202**, 225

rough joint-fir, 6, **128-129**

Rudbeck, Olaus, 65

rudbeckia, 2, **65–66**

Rudbeckia hirta, 2, **65–66**

Rumex crispus, **131–132**

Rumex hymenosepalus, 132, 134

Runyon, Robert, 24

Runyon's nipple cactus, **24-25**

Russian thistle, 5, **245–246**, 284

sacahuiste, **267–268**

sage, autumn, **204–205**

sage, blue, **145**

sage, mealy, **144**

sage, prairie, **123–124**, 133

sage, Texas, **145**

sage, Texas purple, **171**, 183

sagebrush, sand, **123–124**, 133

sagebrush, silver, **123–124**, 133

salad, 62, 65, 68, 73, 76, 97, 98, 100, 137, 144, 156, 160, 164, 166, 167, 213, 231, 237, 250, 252, 266, 270

salsify, yellow, **72–73**, 113

Salsola iberica, 5, **245–246**, 284

Salsola tragus, 5, **245–246**, 284

salt cedar, 19, **222–223**, 227

Salvia farinacea, **144**

Salvia greggii, **204–205**

Salvia texana, **145**

sand bean, **199–200**, 225

sand lily, **264–265**

sand palafoxia, **191–192**, 225

sand penstemon, **219–220**

sand sagebrush, **123–124**, 133

sandbur, prairie, **203**, 225